BASIC ENGINEERING SCIENCE

Dr. N Raghavendra
Department of Mechanical Engineering
B.N.M Institute of Technology, Bangalore-560070

Dr. D Shivalingappa
Department of Mechanical Engineering
B.N.M Institute of Technology, Bangalore-560070

Mrs. K. J. Madhushree
Department of Mechanical Engineering
B.N.M Institute of Technology, Bangalore-560070

Mr. Manu A.S
Department of Mechanical Engineering
B.N.M Institute of Technology, Bangalore-560070

TANEESHA PUBLISHERS

Title : Basic Engineering Science

Authors : Dr. N Raghavendra, Dr. D Shivalingappa, Mrs. K. J. Madhushree, Mr. Manu A.S

Edition : First (February, 2025)

ISBN : 9789348037565

Published by

TANEESHA PUBLISHERS | *A Venture by -* **PRACHI DIGITAL PUBLICATION**

Regd. Add.: 254, Khuriyakhatta No. 10, Bindukhatta, Lalkuan, Nainital - 262402, Uttarakhand, India
Website : www.taneeshapublishers.in
E-mail : taneeshapublishers@gmail.com
Phone : +91 8454 812712, +91 8057 812712

Printed by :
Manipal Technologies Limited, Bengaluru - 560001, Karnataka

Contents

Chapter -1

Energy Sources

1.1. Wind Energy

Introduction

Wind energy is an indirect form of solar energy. Winds are caused by the uneven heating of the atmosphere by the sun, creating temperature, density and pressure differences. It is estimated that 2 per cent of all solar radiation falling on the face of the earth is converted into kinetic energy of the atmosphere, 30 percent of which occurs in the lowest 1000 m of elevation. It is a clean, cheap, and eco-friendly renewable source. The main disadvantages of wind energy are: it is a dispersed, erratic and location-specific source. Wind energy is harnessed as mechanical energy with the help of a wind turbine. Moderate to high-speed winds, typically from 5 m/s to about 25 m/s are considered favorable for most wind turbines.

Globally, wind energy has become a mainstream energy source and an important player in the world's energy markets, and it now contributes to the energy mix in more than 70 countries across the globe.

Major factors that have led to accelerated development of the wind power are as follows:

(i) Availability of high-strength fiber composites for constructing large low cost rotor blades

(ii) Falling prices of power electronics

(iii) Variable speed operation of electrical generators to capture maximum energy

(iv) Improved plant operation, pushing the availability up to 95 per cent

(v) Economy of scale, as the turbines and plants are getting larger in size

(vi) Accumulated field experience (the learning curve effect) improving the capacity factor

(vii) Short energy payback (or energy recovery) period of about one year

Planetary and Local Winds

Winds can be broadly classified as planetary and local. Planetary winds are caused by greater solar heating of earth's surface near the equator than near the northern or southern poles. This causes warm tropical air to rise and flow through the upper atmosphere toward the poles and cold air from the poles to flow back to the equator nearer to the earth's surface. The direction of motion of planetary winds with respect to the earth is affected by the rotation of the earth. The warm air moving toward the poles in the upper atmosphere assumes an easterly direction (in both the northern and southern hemispheres) that results in the prevailing westerlies. (Winds are named according to the direction they come from) At the same time, the inertia of cool air moving toward the equator nearer the earth's surface causes it to turn west, resulting in the north east trade winds in the northern hemisphere and the south east trade winds in the southern hemisphere.

The western motion toward the equator and the eastern motion toward the poles result in large counterclockwise circulation of air around the low-pressure areas in the northern hemisphere and clockwise circulation in the southern hemisphere. The westerlies control events between 30° and 60° latitudes. Because the earth's axis is inclined to its orbital plane around the sun, seasonal variations in the heat received from the sun result in seasonal variations in the velocity and direction of the wind from the general flow pattern described above.

Local winds are caused by two mechanisms. The first is differential heating of land and water. Solar insolation during the day is readily converted to sensible energy of the land surface but is partly absorbed in layers below the water surface and partly consume in evaporating some of that water. The land mass becomes hotter than the water, which causes the air above land to heat up and become warmer than the air above water. The warmer lighter air above the land rises, and the cooler heavier air above the water moves in to replace it. This is the mechanism of shore breezes. At night, the direction of breezes is reversed because the land mass cools more rapidly than the water, assuming a clear sky. The second mechanism of local winds is caused by hills and mountain sides. The air above the slopes heats up during the day and cools down at night, more rapidly than the air above the low lands. This causes heated air during the day to rise along the slopes and relatively cool heavy air to flow down at night.

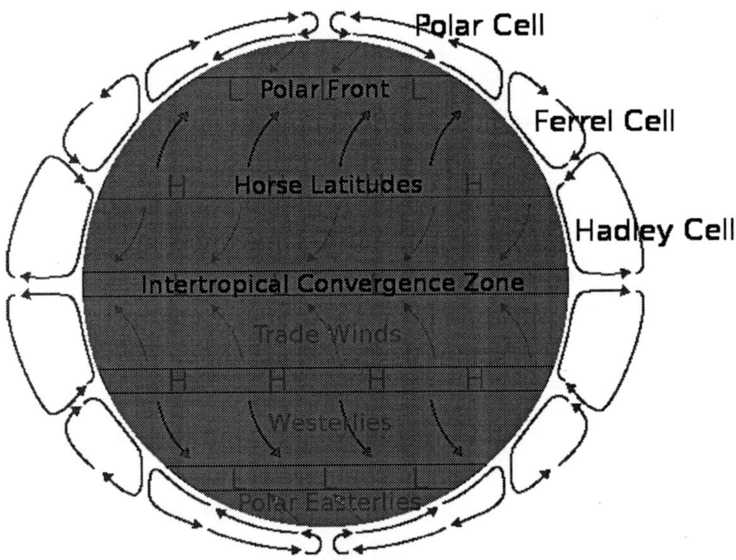

Figure 1.1 Atmospheric Circulation

Wind Data

To be able to understand and predict the performance of wind turbines it is essential to have some knowledge of the behavior and structure of wind. This is also required for proper design of supporting structure to carry the imposed load safely and at an economically viable cost. The Beaufort scale, a wind speed classification, gives a description of the effect of the wind.

Table 1.1 Wind Description Based on Beaufort Number

Beaufort number	Wind Speed		Observable Effects	Wind Description
	m/s	km/h		
0	0.0-0.4	0.0-1.6	Smoke rises vertically	Calm
1	0.4-1.8	1.6-6	Smoke drifts but vanes unaffected	Light
2	1.8-3.6	6-13	Leaves move slightly but vanes unaffected	Light
3	3.6-5.8	13-21	Leaves in motion, flags begin to extend	Light
4	5.8-8.5	21-31	Small branches move, dust raised, pages of book loosen	Moderate
5	8.5-11	31-40	Small trees sway, wind noticeable	Fresh
6	11-14	40-51	Large branches sway, telephone lines whistle	Strong

7	14-17	51-63	Whole tree in motion	Strong
8	17-21	63-76	Twigs break off, walking difficult	Gale
9	21-25	76-88	Slight structural (e.g. chimneys) damage	Gale
10	25-29	88-103	Trees uprooted, much structural damage	Strong Gale
11	29-34	103-121	Widespread damage	Strong Gale
12	>34	>121	Disastrous conditions, countryside devastated, only occurs in tropical cyclones	Hurricane

Wind speed is measured by anemometer and wind direction is measured by wind vane attached to a direction indicator.

According to World Meteorological Organization (WMO) recommendation, wind speed measurement should be made at an effective height of 10 m above the ground. An elegant method of describing average wind speed, duration and direction on a single graph is the wind rose. It depicts the compass bearing from which the wind comes (all 16 directions) along with average wind speed and duration in a year. The lengths of bars represent the percentage of duration.

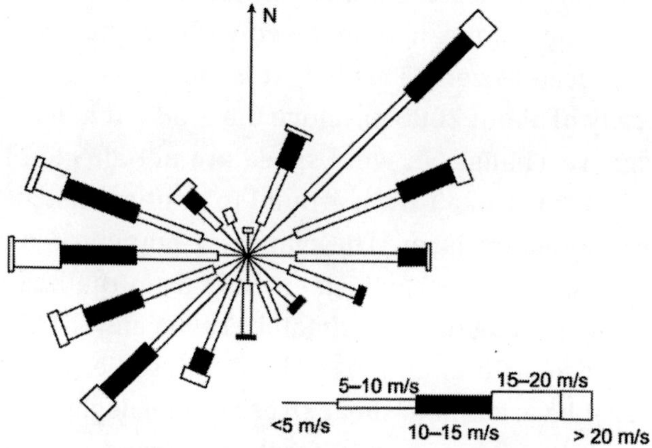

Figure 1.2 A typical Wind Rose

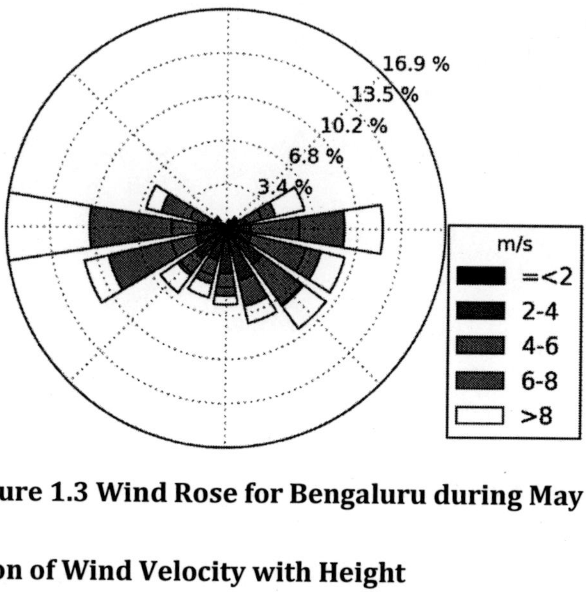

Figure 1.3 Wind Rose for Bengaluru during May 2016

Variation of Wind Velocity with Height

At the earth surface wind speed is always zero. It increases with height above the ground. The wind near the earth surface is retarded by surface roughness. The rate of change of wind speed with height is called wind shear. The lower layers of the air retard those above them, resulting in change in mean wind speed with height, until the shear forces are reduced to zero. This height is called the gradient height and is typically of about 2000 m, above the gradient height, known as free atmosphere, changes in wind speed are not affected by ground conditions. The layer of air from ground to gradient height is known as planetary boundary layer. The planetary boundary layer mainly consists of (i) surface layer, which extends from the height of local obstructions to a height of approximately 100 m and (ii) Ekman layer, which starts from 100 m and extends up to gradient height. In the surface layer the variation of shear stress can be neglected and mean wind speed with height can be represented by Prandtl logarithmic law model:

$$u_z = V \ln\left(\frac{z-d}{z_o}\right)$$

where, V is characteristic speed, d is zero plane displacement, its magnitude is a little less than the height of local obstructions, z_o is roughness length, $(z_o + d)$ is the height of local obstructions.

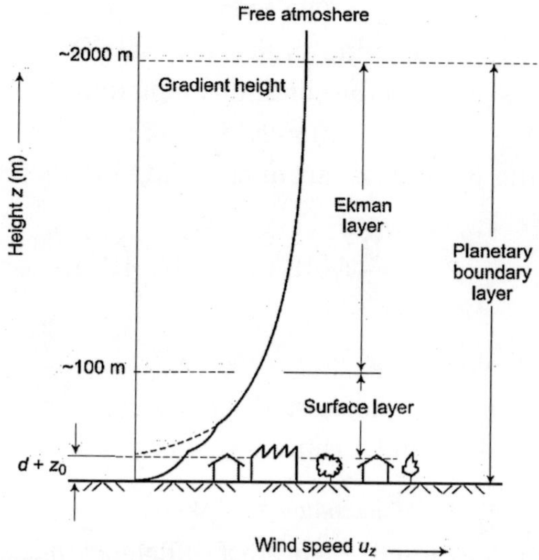

Figure 1.4 Variation of Wind speed with height

It is very important then to place the wind turbine well above the height of local obstructions so that the turbine disk receives a strong uniform wind flux across its area without erratic fluctuations.

As stated earlier, the standard wind speed measurements are often taken at a height of 10 m from ground but wind turbines often operate at a height above this. A simple empirical power law model can be used to estimate wind speed u_z at a height z relative to that available at standard reference height H.

$$u_z = u_H \left(\frac{z}{H}\right)^{\alpha}$$

where u_H is mean wind speed at reference height H (usually 10 m), α depends on surface roughness and the range of height being covered.

Principles of Wind Power

Total Power

The total power of a wind stream is equal to the rate of the incoming kinetic energy of that stream KE_i, or

$$P_{tot} = \frac{\dot{m}V_i^2}{2}$$

$V_i$is the velocity of incoming air,$\dot{m}$ is the mass flow rate of air. Mass flow rate of air is given by the continuity equation;

$$\dot{m} = \rho A\, V_i$$

where ρ is the density of incoming wind, A is the cross sectional area of stream,

$$P_{tot} = \frac{1}{2}(\rho A V_i)V_i^2 = \frac{1}{2}(\rho A V_i^3)$$

Maximum Power

Total Power cannot be converted completely to mechanical power. Maximum power for wind turbine is given by;

$$P_{maximum} = \frac{8}{27}\rho A V_i^3$$

The ideal or maximum, theoretical efficiency $\eta_{\{max\}}$ (also called power coefficient) is given by;

$$\eta_{\{max\}} = \frac{P_{\{max\}}}{P_{tot}} = \frac{8}{27} = 0.5926$$

In other words, a wind turbine is capable of converting no more than approx.. 60% of total power of wind to useful power.

Actual Power

There are many factors that influence the power extraction by a wind turbine like blade tip ratio, blade inlet angle etc. which limits the real efficiency to 50 to 70% of ideal efficiency. The real efficiency η is thus expressed as the actual power produced to total power.

$$P = \eta P_{tot}$$

Forces on the Blades

There are two types of forces operating on the blades of a propeller type wind turbine. They are circumferential forces in the direction of wheel rotation resulting in torque and the axial forces in the direction of wind stream that provide an axial thrust (counteracted by proper arrangement)

Torque,

$$T = \frac{P}{\omega} = \frac{P}{\pi DN}$$

where P is the power extracted, D is the diameter of the wheel and N is the rotational speed of the wheel

The axial force or axial thrust is given by;

$$F_x = \rho A \left(\frac{V_i^2 - V_e^2}{2} \right)$$

Where V_e is exit wind velocity.

Wind Turbine

Types and Classification

Wind turbines are broadly classified into two categories. When the axis of rotation is parallel to the air stream (i.e. horizontal), the turbine is said to be a Horizontal Axis Wind Turbine (HAWT), and when it is perpendicular to the air stream (i.e. vertical), it is said to be a Vertical Axis Wind Turbine (VAWT). The size of the rotor and its speed depends on rating of the turbine. Some of the features of HAWT estimated at wind speed of 12 m/s and $\eta = 30\%$

Horizontal Axis Wind Turbine (HAWT)

Main Components

The constructional details of most common, three-blade rotor, horizontal axis wind turbine are shown in Fig.1.5. Main parts are as follows:

(a) Turbine Blades: Turbine blades are made of high-density wood or glass fiber and epoxy composites. They have airfoil type cross-section. The blades are slightly twisted from the outer tip to the root to reduce the tendency to stall. In addition to centrifugal force and fatigue due to continuous vibrations there are many extraneous forces arising from wind turbulence, gust, gravitational forces and directional changes in the wind, etc. All these factors are to be taken care off at the designing stage. Diameter of a typical, MW range, modern rotor may be of the order of 100 m.

(b) Hub: The central solid portion of the rotor wheel is known as hub. All blades ate attached to the hub. Mechanism for pitch angle control is also provided inside the hub.

(c) Nacelle: The term nacelle is derived from the name for housing containing the engines of an aircraft. The rotor is attached to nacelle, mounted at the top of a tower. It contains rotor brakes, gearbox, generator and electrical switchgear and control. Brakes are used to stop the rotor when power generation is not desired. Gearbox steps up the shaft rpm to suit the generator. Protection and control functions are provided by switchgear and control block. The generated electrical power is conducted to ground terminals through a cable.

(d) Yaw Control Mechanism: The mechanism to adjust the nacelle around vertical axis to keep it facing the wind is provided at the base of nacelle.

(e) Tower: Tower supports nacelle and rotor. For medium and large sized turbines, the tower is slightly taller than the rotor diameter. In case of small sized turbine, the tower is much larger than the rotor diameter as the air is erratic at lower heights. Both steel and concrete towers are being used. The construction can be either tubular or lattice type. The tower vibrations and resulting fatigue cycles under wind speed fluctuations are avoided by careful design. This requires avoidance of all resonance frequencies of tower, the rotor and the nacelle from the wind fluctuation frequencies.

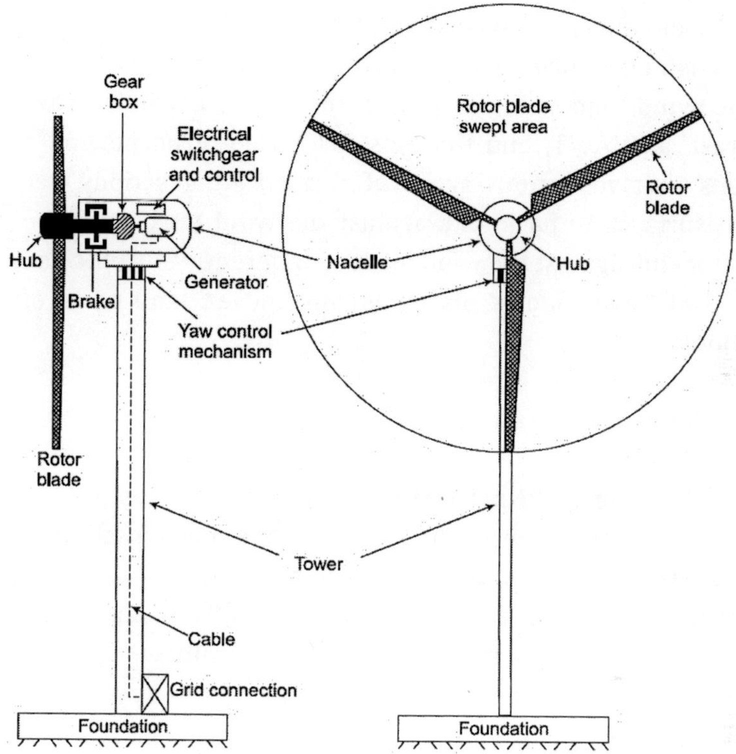

Figure 1.5 Horizontal Axis Wind Turbine

Advantages of Horizontal-axis Wind Turbines

1. Stability: Blades are to the side of the turbines' centre of gravity, helping stability.

2. Ability to wing warp: This gives the turbine blades the best angle of attack.

3. Ability to pitch the rotor blades in a storm to minimize damage: Variable blade pitch, which gives the turbine blades the optimum angle of attack. Allowing the angle of attack to be remotely adjusted gives great control, so the turbine collects the maximum amount of wind energy for the time of day and season.

4. Tall tower allows access to strong wind in sites with wind shear:

In some wind shear sites, every 10m up, the wind speed can increase by 20% and the power output by 34%.

5. High efficiency: Since the blades always move perpendicularly to the wind, and receives power through the whole rotation. In contrast, all VAWT, and the most proposed airbornewind turbine designs involve various types of reciprocating actions, requiring airfoil surfaces to backtrack against the wind for part of the cycle. Backtracking against the wind leads to inherently low efficiency.

6. Tall tower allows placement on uneven land or in offshore locations.

7. Can be sited in forest above tree-line.

8. Most are self-starting.

Disadvantages of Horizontal-axis Wind Turbines

1. It is difficult to transport (20% of equipment costs) and install. Tall masts and blades are more difficult to transport and install. Transportation and installation can now cost 20% of equipment costs. Further, it requires tall cranes and skilled operators.

2. Strong tower construction is required to support the heavy blades, gearbox, and generator.

3. Effect radar in proximity reflections from tall HAWTs may affect side lobes of radar installation screating signal clutter, although filtering can suppress it.

4. Difficult to maintain.

5. Fatigue issues leading to blade failure.

6. They require an additional yaw control mechanism to turn the blades toward the wind.

Vertical Axis Wind Turbine (VAWT)

VAWTs are in the development stage and many models are undergoing field trial. Main attractions of a VAWT are:

(i) It can accept wind from any direction, eliminating the need of yaw control.

(ii) Gearbox, generator etc. are located at the ground, thus eliminating the heavy nacelle at the top of the tower. This simplifies the design and installation of the whole structure, including tower.

(iii) The inspection and maintenance also gets easier and

(iv) It also reduces the overall cost.

The constructional details of a vertical axis wind turbine (Darrieus type rotor) are shown in Fig. 1.6. The details of main components are as follows:

(a) Tower (or Rotor Shaft): The tower is a hollow vertical rotor shaft, which rotates freely about vertical axis between top and bottom bearings. It is installed above a support structure. In the absence of any load at the top, a very strong tower is not required, which greatly simplifies its design. The upper part of the tower is supported by guy ropes. The height of the tower of a large turbine is around 100 m.

(b) Blades: It has two or three thin, curved blades shaped like an eggbeater in profile, with blades curved in a form that minimizes the bending stress caused by centrifugal forces-the so-called 'Troposkien' profile. The blades have airfoil cross section with constant chord length. The pitch of the blades cannot be changed. The diameter of the rotor is slightly less than the tower height.

(c) Support Structure: Support structure is provided at the ground to support the weight of the rotor. Gearbox, generator, brakes, electrical switchgear and controls are housed within this structure.

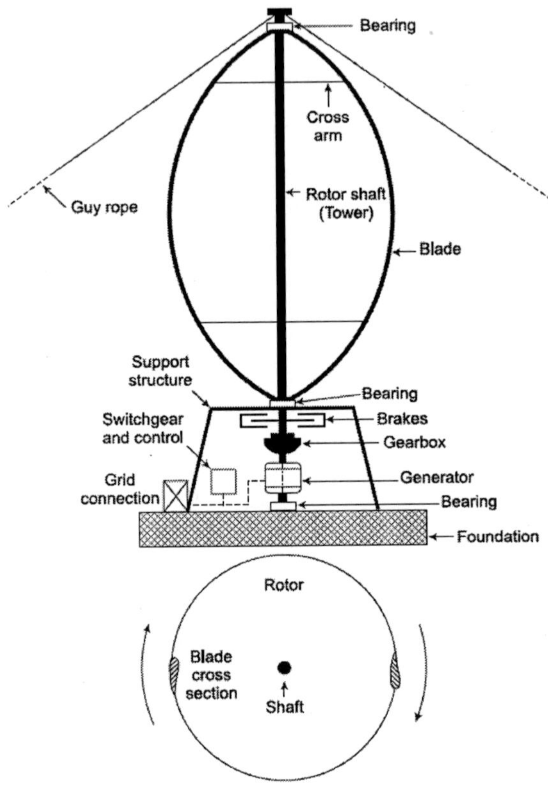

Figure 1.6 Darrieus type Vertical Axis Wind Turbine

1.2. Tidal Energy

The movement of the ocean's waves, tides, and currents carries kinetic energy that can be harnessed and converted into electricity to power homes, buildings, and cities. This movement occurs naturally when waves crash against coastlines and tidal currents ebb and flow.

Tides are created because of the gravitational and kinematic effects due to the position and the combined motion of the Earth, the Moon and the Sun. The Moon, even though it has much smaller mass than the Sun, plays a more important role in the creation of tides because it is significantly closer than the Sun. The mass of the Moon "pulls" the ocean water masses in its direction and creates a "bulge" on the surface of the ocean. The effect of the Moon's pull is modified by the effect of the Sun, as shown in Fig. 1.7. When the Earth, the Moon and the sun are aligned, the tides are amplified and are called spring tides. When the three bodies form a right-angle triangle the tidal effect is modulated and we have the neap tides. There are two periods of spring tides and two of neap tides during every Moon cycle, which lasts for approximately 29.5 days.

A simple analysis of the effects of the gravitational potentials of the Earth and the Moon on the surface of the Earth's ocean may be made with the help of Fig. 1.8., which shows the combined effect of the Moon's gravity on the surface of the ocean and the rotation of the Earth around its axis. When we consider an equipotential surface around the surface of the Earth and perform the force balance resulting from the gravitational forces of the Moon and the Earth on a unit mass of water, at the surface of the ocean, we will obtain the following approximate expression for the height of the ocean "bulge" at any point on the surface of the ocean:

$$h(\theta) = \frac{r^4 m}{M d^3}\left(\frac{3}{2}\cos^2\theta - \frac{1}{2}\right)$$

where θ is the angle formed between the line that connects the centers of the Earth and the Moon and the line that connects the center of the Earth with the point under consideration. M is the mass of the Earth; m is the mass of the Moon; r is the average radius of the Earth, which is approximately 6,387 km; and d is the Moon to Earth distance, which is approximately 384,420 km. Given that $m/M =$ 0.0123, the equation reveals that the maximum bulge on the surface of the oceans is 0.36 m. This value is close to the globally-averaged height of the tides on the oceans, which is measured to be approximately 0.5 m. The range of the tides, H, is the difference between the maximum and the minimum heights of the water.

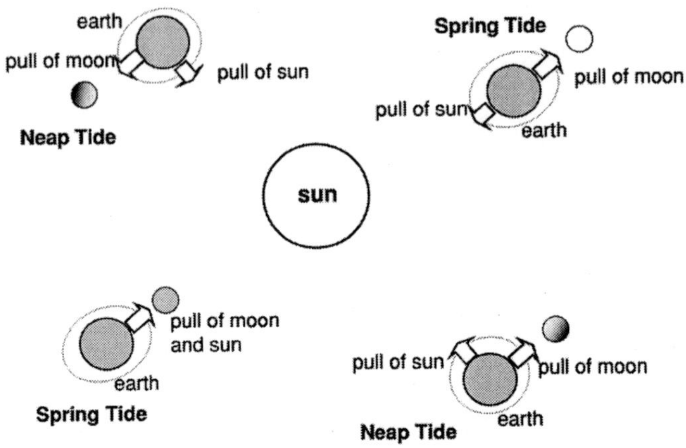

Figure 1.7 Spring and Neap Tides

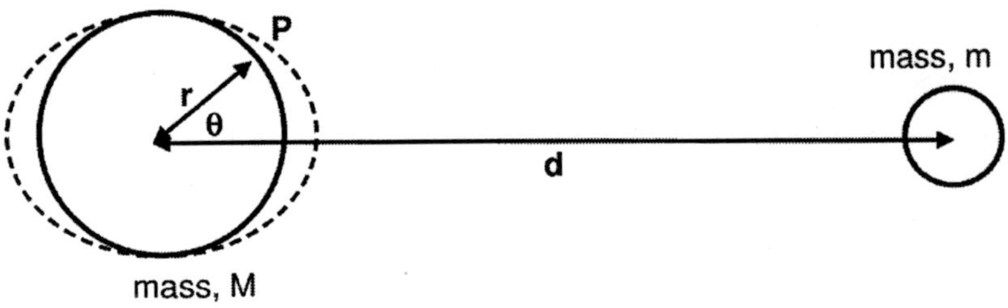

Figure 1.8 Gravitational Interaction of Earth and Moon

The average range of the surface of the oceans is twice the maximum height, or approximately 1 *m*.

The bulges on the tidal deformations on the surface of the ocean form the tidal waves, which are shallow, long wavelength waves that move on the surface of the Earth's ocean. At any time, there are two highs and two lows on the surface of the ocean.

The Earth rotates around its axis with a period of 24 h, and the apparent motion of the moon is a rotation around the Earth with a period of 24 h and 50 min. This implies that the tidal waves propagate on the surface of the oceans with a period, which is half the apparent period of rotation of the Moon. Therefore, the period of the ocean tides is 12 h 25 min or approximately 45,000 s.

According to elementary wave theory, typical speeds of propagation of the tidal waves are close to 200 m/s. From the wave equation:

$$c = \lambda \nu = \frac{\lambda}{T}$$

one may conclude that the tidal wavelengths are approximately 9,000 km. This distance is greater than the width of the Atlantic Ocean and of the same order of magnitude as the maximum width of the Pacific Ocean.

The presence of the continental shelves, the variable ocean depth and the irregularity of the coastlines interfere significantly with tidal waves and cause local distortions and resonances. The resonances create tidal waves with different amplitudes locally, which are significantly higher than the global average range of 1 m. These are most common in bays, gulfs and estuaries. Tidal ranges between 5 and 10 m have been observed in several narrow bays and estuaries. Several of these high ranges are listed in Table 11.3. These ranges reveal a few of the locations on the Earth with high potential for electricity production from tidal power.

Table 1.2 shows that the amount of tidal power, which is available for conversion to electricity, is very high. The average power available in several of the locations listed in this Table is equal to that produced by tens

of conventional fossil fuel plants. For example, the power available in the Cobequid area alone is equivalent to 31% of the total electricity that was consumed in Canada in 2009.Similarly, the full utilization of the tidal power resources in France and England has the capacity to provide between 25 and 30% of the electricity consumption in the two countries. Clearly, tidal power if developed and well utilized may provide a high percentage of the electric energy demand and will prove to be an excellent alternative to fossil fuels.

Table 1.2 Locations with high potential for tidal energy conversion (one pool system)

Location	Country	Range, m	Average Power (MW)	Annual Energy, (GWh)
Cobequid	Canada	10.7	20000	175200
Passmaquaddy	Canada	5.5	1800	15768
Sevem river	England	9.8	1680	14717
La Rance	France	8.4	349	3956
Mont St. Michel	France	8.4	9700	84972
Kimberlay	Australia	6.4	630	5519
Mesen	Russia	6.6	14000	122640
Khambat	India	7.0	7000	61320

Tidal Energy Utilization

Main tidal energy conversion schemes are: (i) single basin: single effect, (ii) single basin: double effect, (iii) two basin: linked basin (iv) two basin: paired basin and (v) tidal flow (or tidal current) schemes.

(a) Single Basin: Single Effect Scheme Single basin scheme has only one basin as shown in Fig. 1.9. In single effect scheme, power is generated during either filling or emptying the basin. Two types of operation cycles are possible. In ebb generation cycle operation, the sluice way is opened to fill the basin during high tide. Once filled the impounded water is held till the receding cycle creates a suitable head. Water is now allowed to flow through the turbine coupled to generator till the rising tide reduces the head to the minimum operating point. The flow is held till the next generating cycle.

The sequence of events is illustrated in Fig. 1.10. This cycle is repeated and power is generated intermittently. In flood generation cycle operation, the sequences are altered to generate power during filling operation of the basin. However, the sloping nature of the basin shores usually makes ebb generation the more productive method.

Increased output can also be obtained by pumping (using turbines as pumps) during high tide to increase the basin level and therefore the generation head. The energy required for pumping must be borrowed and repaid. As the pumping is done against a small head at high tide, whereas the same water is released through the turbine during low tide at a great head, producing a net energy gain.

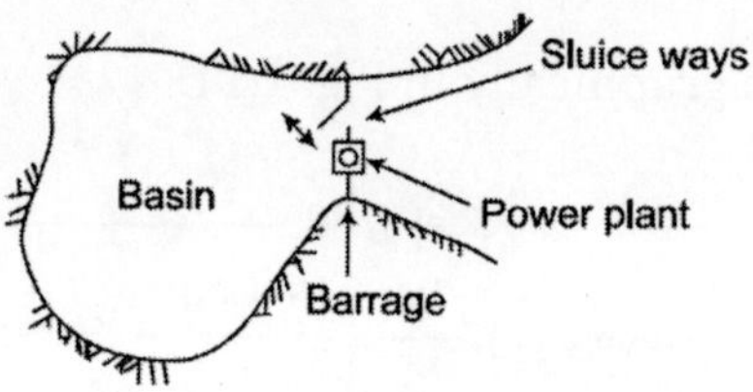

Figure 1.9 Layout of Single Basin Single Effect Scheme

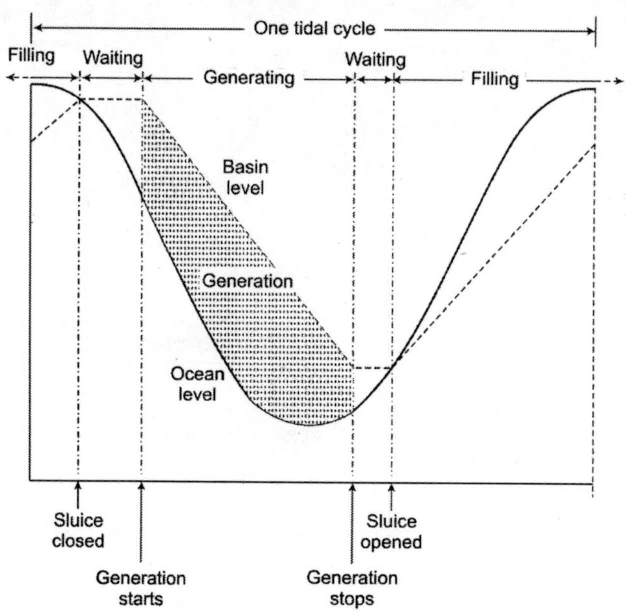

Figure 1.10 Sequence of Operations (Single basin Single Effect System)

(b) Single Basin: Double Effect Scheme In single basin, double effect scheme, power is generated on both flood and ebb. Two-way (reversible) hydraulic turbines are used. Pumping may also be used to increase the output. The routine is as follows:

1. inward sluicing to fill the basin
2. holding period
3. ebb generation
4. outward sluicing to empty the basin
5. holding period
6. flood generation

The routine is shown graphically in Fig. 1.11.

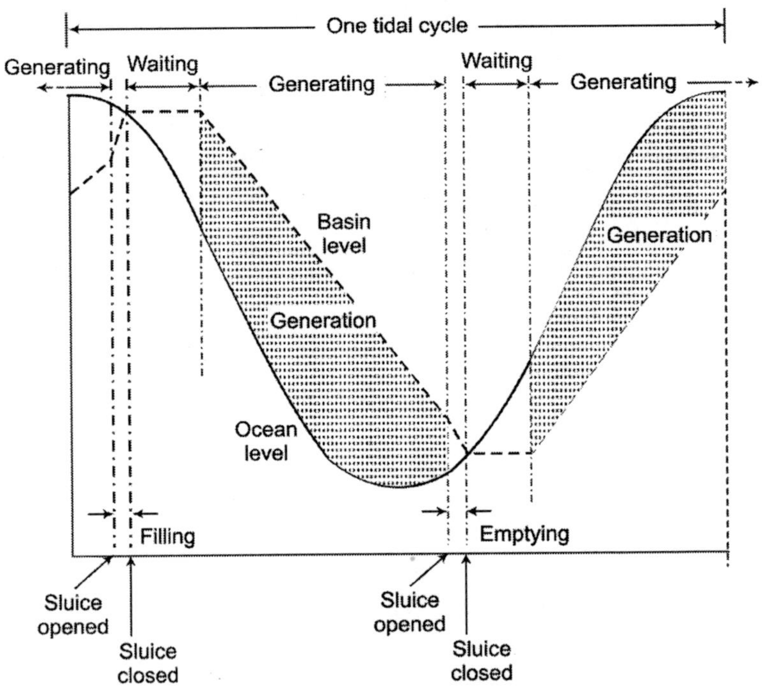

Figure 1.11 Sequence of Operations (Single Basin Double Effect Scheme)

(c) Two Basin: Linked Basin Scheme In order to maintain continuity of power supply linked and paired basins schemes are used. Linked basin scheme consists of two basins, one toped up at high tide and the other emptied at low tide. Thus a permanent head is created between the two basins. Water flows through turbine from high basin to low basin. The layout of such a scheme is shown in Fig. 1.12.

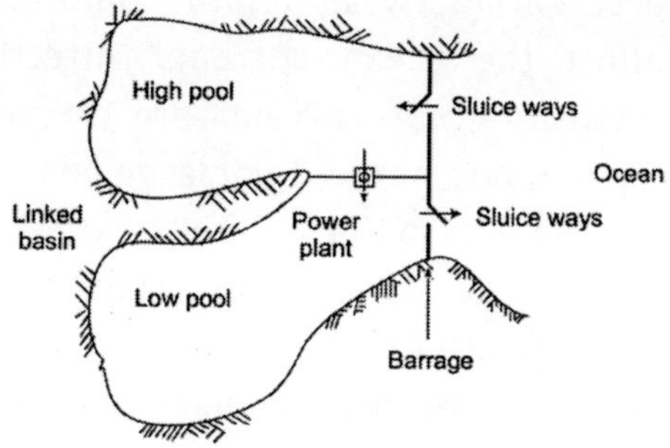

Figure 1.12 Layout of Linked Basin Scheme

(d) Two BasinPaired Basin Scheme A paired basin consists essentially of two single basin schemes. One scheme generates on flood and the other on ebb cycle. The output is almost, but not quite continuous. Two basin schemes are generally found to be economically inferior to single basin schemes.

(e) Tidal Current Schemes In order to explore the possibility of dispensing with the dam, its associated cost and environmental impacts, extraction of energy from tidal currents have been considered. Strong tidal currents (as high as about 5 m/s) are found in shallow seas, particularly where natural constrictions exist, such as around headlands or between islands. The practical devices for tidal flow power will be similar to that used in river flow power systems. The theory of tidal flow power is similar to wind power with the advantage of predictable velocities of high-density fluid. However, in this case the fluid velocities are much less as compared to wind and the operation is carried out in marine environment. The generation will be intermittent and total power available may not be very large. The capital cost per unit of power produced appears to be high. As the tidal flow power lags about $\pi/2$ behind tidal range power from a single basin, the two systems could be complementary.

1.3. Wave Energy Conversion Devices

Permanent ocean currents occur in several parts of the sea and are generated by a variety of forces including: the Coriolis force; the predominant surface wind; temperature and salinity gradients; the predominant tides. Variability in depth contours and shoreline configurations affect the ocean currents' direction and power potential. High intensity ocean currents are frequently met in the waterways between islands, at the tip of large peninsulas and capes, for example off Cape Hateras in North Carolina, U.S.A. or at waterways that connect oceans, as for example the Straight of Magellan at the southernmost point of South America.

An ocean current is in all respects similar to a wind current. As with the wind power, the available power of ocean currents is given by the expression:

$$W = \frac{1}{2} A\, \rho V^3$$

The power production limitation on wind power, or Betz's Law is also applicable to the ocean currents. The maximum power that may be produced from a water turbine, which is placed within the current, would be:

$$W_{maximum} = \frac{8}{27} \rho A\, V_i^3$$

A significant difference between wind and ocean currents is that the density of the water is approximately 800 times higher than that of air. This implies that significantly more power may be produced by water currents. For example a small ocean current turbine with 3 m blade diameter would produce a maximum of almost 152 kW of electric power. Such an engine would only produce 0.20 kW from the wind. As a result, sea current turbines may be significantly smaller and more compact than their wind counterparts.

Main advantages of wave power are: (i) the availability of large energy fluxes (as will be seen in the next section) and (ii) predictability of wave conditions over periods of days.

Difficulties in the development of wave power are encountered mainly due to following reasons:

(i) Irregularity of wave patterns in amplitude, phase and direction, which makes it difficult to extract power efficiently

(ii) The power extraction system is exposed to occasional extreme stormy conditions. During unusual extreme conditions, once in several years, the wave amplitude may reach as high as 10 times normal value and the associated power about 100 times the normal values. The devastating tsunami near the coast of Indonesia towards the end of December 2004 is the most recent example in hand. Allowing for this is expensive and would reduce normal efficiency of power extraction.

(iii) Peak power of deep-water waves is available in open sea, where it is difficult to construct, operate and maintain a system and transmit power to the shore.

(iv) The slow and irregular motion of wave is required to be coupled to electrical generator requiring high and constant speed motion.

Energy in the waves is harnessed basically in the form of mechanical energy using wave energy converters, also known as wave devices or wave machines. A wave device may be placed in the ocean in various possible situations and locations. The fluctuating mechanical energy obtained is modified/smoothed out to drive a generator. Wave energy devices may be classified in various ways. Depending upon the location of these devices, it can be classified as **(i)** off shore or deep-water (water depth of 40 more) devices and **(ii)** shoreline devices. The availability of wave power at deep ocean sites is three to eight times that of adjacent coastal sites. However, the cost of construction/installation, operation and power transmission is quite large. Shoreline devices on the other hand have the advantage of relatively easier maintenance and installation and they do not require deep-water moorings and long underwater cables. Depending upon the position with respect to sea level they are classified as:(i) floating, (ii) submerged and (iii) partly submerged devices. Submerged devices have the benefit of avoiding the worst storm conditions but have increased difficulties of maintenance.

Depending on the actuating motion used in capturing the wave power, these devices are classified as: (i) heaving float type, (ii) pitching type, (iii) heaving and pitching float type, (iv) oscillating water column type and (v) surge devices. For each of these types several designs have been developed and tested with many more still at the design stage. Some designs will be discussed here:

1. Heaving Float Type

A float (buoy) placed on the surface of water heaves up and down with waves due to rise and fall of water level. The resulting vertical motion is used to operate the piston of an air pump through linkage. The pump may be anchored or moored to seabed. Several float-operated air pumps are used to store energy in compressed air storage. The compressed air is used to generate electricity through air turbine coupled to generator.

Power may also be extracted from a buoy by means of an internally suspended weight, with high inertia but free to swing. Energy is extracted from differential motion of buoy and pendulum. In another version, hydraulic pump is operated by the motion of buoy to raise water to an onshore reservoir and passed through a turbo-generator to generate electricity. In a Swedish design a submerged flexible tube attached to buoy is used as air pump. Motion of buoy stretches the tube, decreasing its volume to provide the pumping action. These designs are shown in Fig. 1.13.

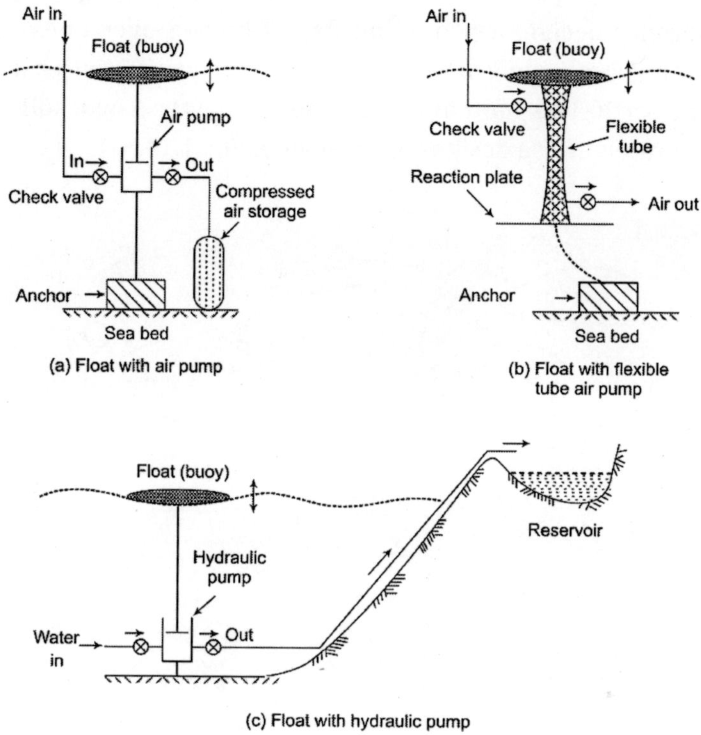

Figure 1.13 Heavy Floating Type Devices

2. Pitching Type

In pitching type devices, the waves strike horizontally on a floating piece or flap causing it to deflect. In one such model several cam shaped floating pieces (known as nodding ducks) are hinged to a common flexible linkage or spine to form nodding duck assembly. A wave entering from left turns the beak of the duck in the direction of motion. The duck swings backward with trough of the wave once the peak has passed over it. It then oscillates about its axis. The shape of the duck is optimized to extract maximum power. The ratchet and wheel mechanism converts the motion of the ducks to common shaft. Power collected by individual duck is thus pooled to drive generator through linkages and gears. Main advantage of this system is that during unusually large wave the duck can flip over and recover again.

In another design, a pendulum flap is hinged over opening of a terminating rectangular box. The action of the waves causes it swing back and forth. This motion is then used to power a hydraulic pump and generator. These designs are shown in Fig. 1.14.

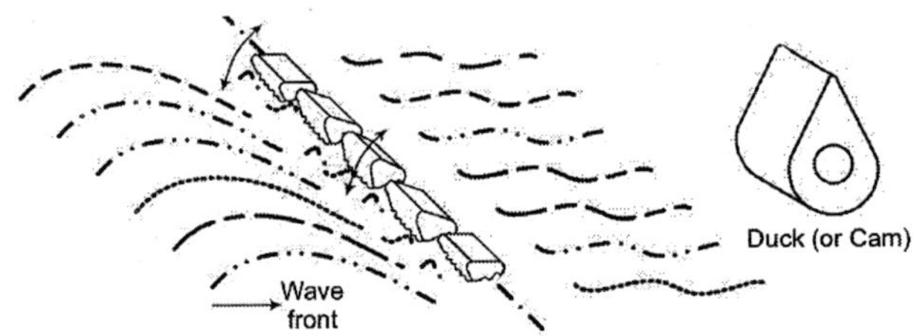

(a) Nodding ducks assembly

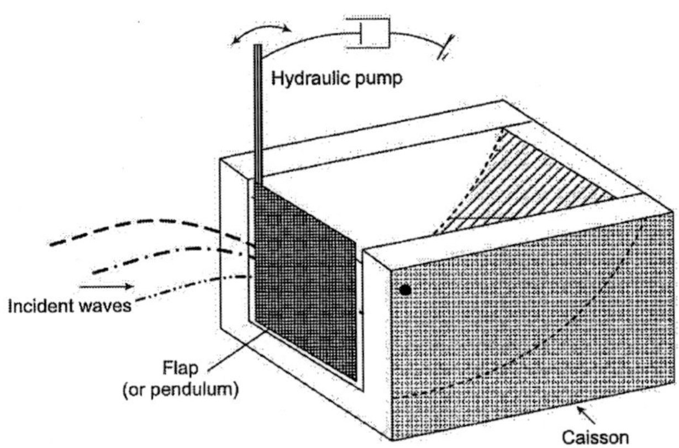

(b) Pendular device

Figure 1.14 Pitching Devices

3. Heaving and Pitching Float Type

Pitching and heaving motion of a float can also be used to extract wave energy. Several designs have been tried. A system consisting of three pontoons, which move relative to each other in the wave, is shown in Fig. 1.15. (a). A damper plate is attached to central pontoon, which ensures that it stays still as the fore and aft pontoon move relative to central pontoon by pitching about the hinges. Energy is extracted by hydraulic pumps attached to hinges. Electricity can be generated via a hydraulic turbine and generator.

The idea of three pontoons system is further modified to form "Pelamis Wave Energy Converter". In this design a number of semi-submerged cylindrical sections (known as "Pelamis") are joined with flexible linkages to follow surface wave profile.

The sections articulate with the movement of the waves, each resisting motion between it and the next section, creating pressurized oil to drive a hydraulic ram which drives a hydraulic motor. The machine is long and narrow (snake-like) and points into the waves; it attenuates the waves, gathering more energy than its narrow profile suggests. Its articulating sections drive internal hydraulic generators (through the use of pumps and accumulators).

A specially shaped float known as Dolphin rides the wave and rolls as well as heaves as the wave passes. The two motions are converted to unidirectional motion by ratchets wheel arrangement and used to operate floating and stationary generators respectively. In another very simple design a large float is used to produce differential motion between two extreme sides of the float. The motion can be used to operate hydraulic pump. These concepts are shown in Fig. 1.15.

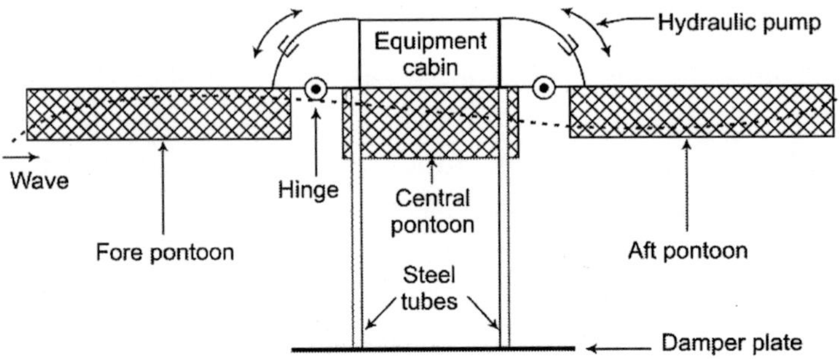

(a) Three pontoon with hydraulic pump

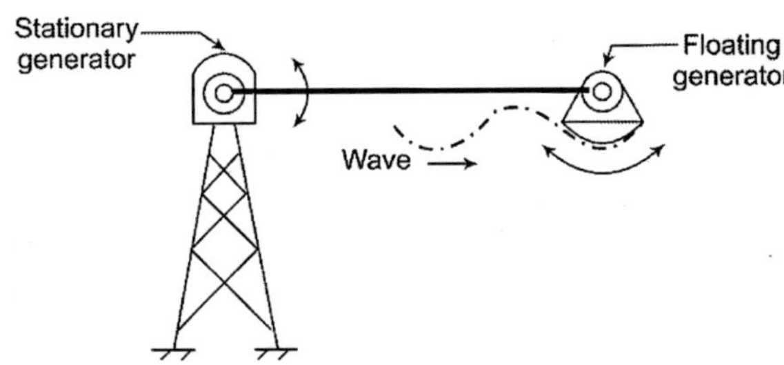

(b) Buoy dolphin

Figure 1.15 Heaving and Pitching Devices

4. Oscillating Water Column Type

This device comprise a partly submerged concrete or steel structure, which has an opening to the sea below the water line, thereby enclosing a column of air above a column of water. The wave impinging on the device causes the water column to rise and fall, which alternately compresses or depressurizes the air column. The air is allowed to flow through a turbine, which drives the generator. The specially developed axial flow Well's turbine, which operates with either directions of airflow, is used to extract energy as shown in Fig. 1.16. Conventional (unidirectional) air turbine with rectifying valves can also be used. Main advantages of oscillating water column device are: (i) the air velocity can be increased by reducing the cross section area of the air channel. Thus the slow wave motion can be coupled to high-speed turbine motion. (ii) generating equipment are kept away from immediate saline water environment.

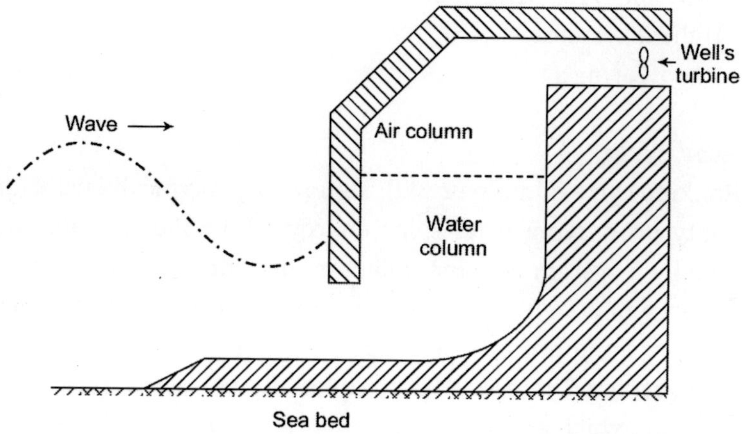

Figure 1.16 Oscillating Water Column Device

1.4. Solar Thermal Energy Conversion

The common characteristic of all Solar-Thermal systems for energy conversion is that they capture the solar radiation as heat and subsequently use this heat primarily for two purposes:

a) The production of electricity via a conventional thermal cycle, and b) Residential space or water heating and the supply of heat for commercial and industrial processes.

For this process to be efficient and economical, it is necessary to collect and concentrate the diffuse solar radiation in an efficient manner to arrive at a reasonably high-temperature heat source. Solar thermal collectors gather the sun's energy and direct it onto the receivers that contain the working fluid. The receivers that can be considered are:

1. Central receivers
2. Dispersed or distributed receivers – can be a) the point-focus type or b) the line-focus type
3. Solar ponds

The conversion systems used in solar thermal electric conversion are of many types. These include:

1. Rankine cycle, using steam or other working fluid
2. Brayton cycle, using helium or air as working fluid
3. Hybrid systems
4. Repowering systems

Power Cycle

A Rankine cycle (simple or with superheat) is typically used for the electricity production from solar energy. The boiler in this cycle is replaced by a system of solar collectors or reflectors that impart the solar energy to water or another working fluid, which undergoes the Rankine cycle and produces electric power in a conventional turbine. Fig. 1.17. depicts the essential components of such a solar-thermal power plant, which are:

1) The pump that circulates the working fluid;

2) The system of solar reflectors and a central receiver that receives the solar energy and raise vapor;

3) The vapor turbine, usually working with steam and

4) The condenser.

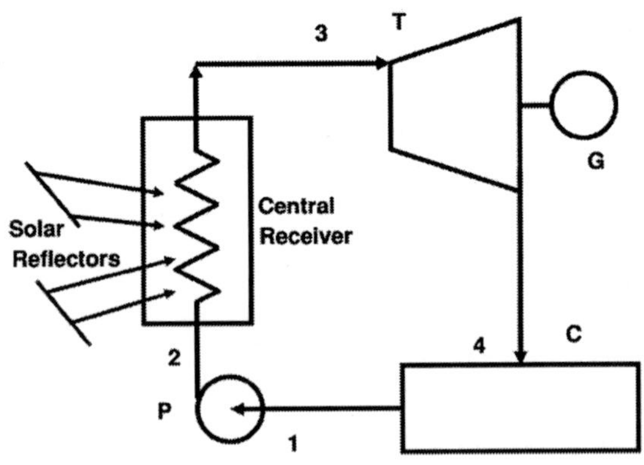

Figure 1.17 Essential Components of a Solar Thermal Power Plant

1.4.2. Solar-Thermal Central Receiver Systems

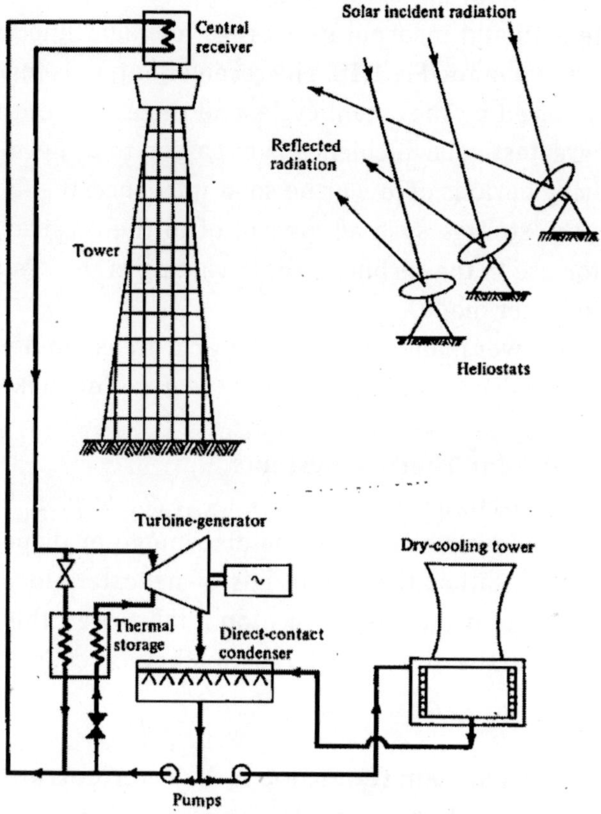

Figure 1.18 Central Receiver-Heliostat Based Solar Power Plant

The central receiver approach to solar-thermal-electric systems uses a large field of reflecting mirrors called heliostats that redirect the sun's energy and concentrate it on a central receiver mounted on top of a tower. The heliostats are individually guided, since they cover a large field, so that each focuses the sun's energy it receives on the central receiver at all hours of sunlight. In the receiver the concentrated solar energy is absorbed by a circulating fluid. The fluid could be water, which vaporizes into steam that is used to drive a turbogenerator in a Rankine cycle, or an intermediate fluid that transports the heat to the steam cycle.

The system should incorporate storage for nighttime and cloudy periods as shown in the Fig.1.18. The receiver output is made greater than that required by the steam cycle, and the excess output during periods of greatest solar incidence is bypassed to a thermal storage system. During periods of low or no solar incidence, the feedwater is shunted to the storage system, instead of to the receiver, where it vaporizes for use in the turbine, proper valving in the system allows operation in either mode.

Since these power plants are located in arid areas, condenser water is cooled using a dry cooling tower which requires no makeup water.

Distributed Solar Thermal Systems

Another approach to solar thermal energy conversion, besides the central receiver tower concept is the distributed or dispersed solar thermal system. Rather than hundreds of heliostats focusing solar energy on a single distant receiver atop a tall tower, this system is characterized by the use of a large number of collectors, called concentrators.

Parabolic Dish System (Point-focus dish concentrator)

A parabolic dish collector is similar in appearance to a large satellite dish as shown in Fig.1.19, but has mirror-like reflectors and an absorber at the focal point. It uses a dual-axis sun tracker.

1. Concentrating solar energy onto a receiver located at the focal point of the dish.

2. The dish structure must track fully the sun to reflect the beam into the thermal receiver.

The receiver absorbs the radiant solar energy, converting it into thermal energy in a circulating fluid. The thermal energy can then either be converted into electricity using an engine–generator coupled directly to the receiver, or it can be transported through pipes to a central power conversion system. Parabolic-dish systems can achieve temperatures in excess of 1,500°C. Because the receivers are distributed throughout a collector field, like parabolic troughs, parabolic dishes are often called distributed receiver systems.

The following are the important advantages of parabolic dish reflectors:

1. They are the most efficient of all collector systems as they are always pointing the sun.

2. They typically have concentration ratio in the range of 500–2,000, and thus are highly efficient at thermal energy absorption and power conversion systems.

3. They have modular collector and receiver units that can either function independently or as part of a large system of dishes.

This type of solar thermal energy concentrator is primarily used with parabolic dish engine, which is usually a Stirling engine coupled to a generator.

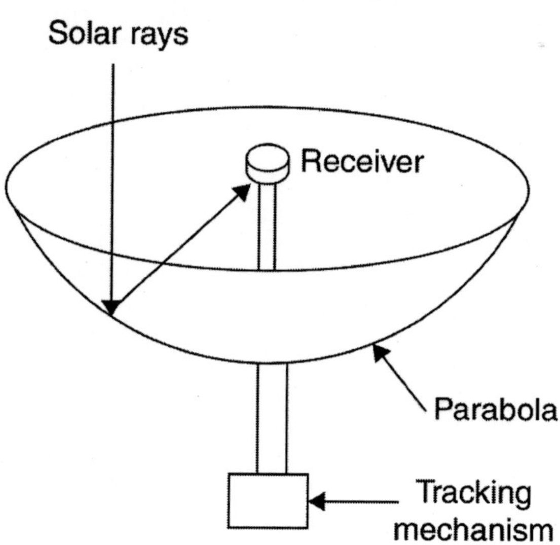

Figure 1.19 Parabolic Dish System

Cylindrical Trough Solar Collectors

Since linear translation does not introduce defocusing of the concentrated radiation, the aperture of a cylindrical trough need not track at all to maintain focus. However, as indicated in fig., a high-rim angle cylindrical trough would have a focal plane not a focal line. Toavoid a dispersed focus, cylindrical troughs would have to be designed with low rim angles inorder to provide an approximate line focus. The advantage of cylindrical mirror geometry is thatit need not track the sun in any direction as long as some means is provided to intercept the moving focus.

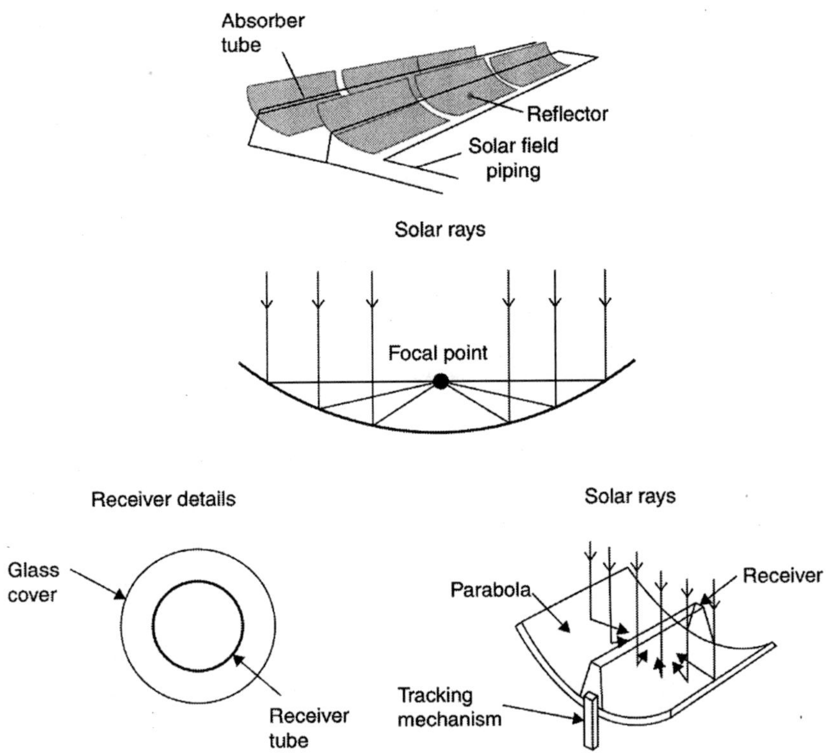

Figure 1.20 Cylindrical Trough Solar Collector

Solar Pond

The concept of the solar ponds combines power production from solar energy and storage. A solar pond is essentially a pool of water, approximately 1 m deep. In a typical pool the water is heated by the Sun and rises to the surface, where it partly evaporates. The evaporation of the water vents a great deal of the pool's water as vapor; does not allow the accumulation of a great deal of energy in the pool; and keeps the average temperature of the water to low levels. Actually, the temperature of a typical pool in the summer is low enough for it to be used for recreational activities. If the natural convection process that brings warmer fluid to the pool surface is suppressed, it is possible for the water temperature at the bottom of the pool to increase significantly by absorbing solar radiation. This may be accomplished in a salt-gradient solar pond, which uses natural salt (NaCl) as well as transparentpartitions to suppress the natural convection in the pool. The depth of the solarpond is approximately 1 m and its bottom and walls are coated with a light absorbing material. The water in the solar pond is usually separated in three layers:

a) The bottom layer is saturated with salt and, therefore, its density is significantly greater than the density of pure water

b) The intermediate layer, which also contains salt at lower concentration, and

c) The upper layer, which is pure water without any salt.

Even though some solar ponds do not have partitions between the three layers,it is advisable to include transparent partitions in order to suppress moleculardiffusion, which will disturb and annihilate the salt gradient.

When sunlight enters the solar pond, it is captured by the absorbing bottom andwalls as well as the water. The temperature of the water close to the bottom rises,but the density of the salty water in this layer is high enough for natural convectionto not occur. The transparent partitions between the three layers create a"greenhouse effect," thus trapping an additional quantity of heat in the two bottomlayers. A great deal of the insolation absorbed remains at the bottom of the pond,where the temperature rises significantly. Temperatures as high as 95°Chave beenrecorded in the bottom layer of well-designed solar ponds. Temperatures of thismagnitude are sufficient to produce vapor in a suitable secondary fluid, such aspentane, butane or a refrigerant, which becomes the working fluid of a Rankinecycle operating with the solar pond as the boiler.

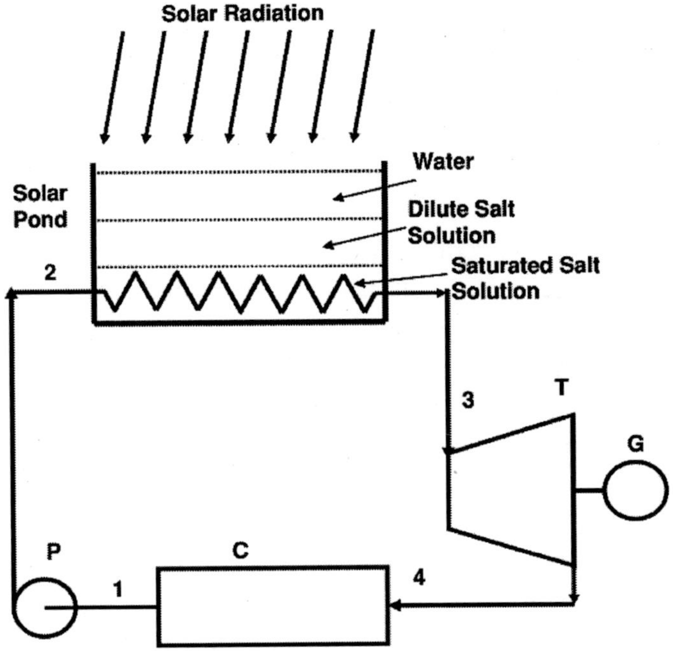

Figure 1.21 Solar Pond based Power Plant

Because the water has very high heat capacity, a solar pond may store a significant amount of energy to be used when the Sun is low in the horizon. The production of electricity from a solar pond may continue even in the evening hours, when there is no natural insolation, thus providing a good alternative to produce power in the evening hours, when the demand is high because of air-conditioning needs. The main advantage of solar ponds is that they serve simultaneously as electric power generating plants as well as storage media.

The disadvantage of solar ponds is that they loose a significant amount of heat to the ground and to the atmospheric air. Because of this, the temperature of the lower zone becomes high enough to produce power for only a limited period of time during the year, typically for a few weeks during the summer. Even during these weeks, the solar pond temperature drops during the nights to levels that do not support power generation. Therefore the energy efficiency of solar ponds defined as the electric energy produced during a year divided by the total incident solar energy is expected to be very low.

Passive Solar Heating: Solar Collectors

The most widespread use of solar energy is the passive heating of buildings and water. While passive heating does not produce electricity or any other form of energy, it helps avoid the use of fossil fuels or electric power for heating purposes and, thus, helps conserve the energy resources of the planet.

Passive solar heating in buildings is caused by an artificial greenhouse effect: Glass and several polymeric materials allow low-wavelength solar radiation to pass through them, but trap high-wavelength radiation. These materials effectively act as high-wavelength radiation filters. An enclosure that is covered by one of the greenhouse materials and faces the Sun "traps" a great deal of the solar energy.

As a consequence, the temperature inside the enclosure will rise. Thermodynamic equilibrium between the enclosure and its surroundings will be established when the heat transfer from the enclosure—via conduction and advection—is equal to the difference of the radiation energy that enters and exits the enclosure. A simple piece of glass placed on the top of an otherwise open water tank, which is exposed to the Sun in the southern European countries, would raise the temperature of the water by 15 – 25°C. This is an adequate temperature for the water to be used in domestic applications. Well-insulated enclosures may reach temperatures 40 – 50°C higher than the ambient. Such enclosures are called solar collectors or simply collectors. The solar collectors are typically placed on the roofs of buildings, facing the prevailing direction of the Sun—south in the northern hemisphere and north in the southern hemisphere—and are used for water and space heating.

A schematic diagram of the cross-section of a solar collector is depicted in Fig.1.22. Because a typical collector is placed on a roof, the solar collector depicted is at an angle with the horizontal. The inside material of the collector is coated with absorbing paint and usually has a dark color. Solar energy enters the top of the collector through the double glass and heats up the enclosed fluid, which is typically water or a solution. Depending on the design of the collector, the temperature of this fluid may reach 70 – 85°C. Another, closed water circuit picks up the heat from the collector and dissipates it inside the building: the water that is circulated through the pipes of this circuit may supply the tank of a water heater and maintain it at the typically required temperature of 40 – 50°C, or it may circulate via space heaters to supply the heat required to maintain a comfortable condition in the building.

Solar collectors may trap direct or diffuse solar radiation. Because they also absorb diffused radiation, they may operate even in the cloudy autumn and winter days when most of the heating for the buildings is needed. Since the heated water may be stored for later use, the solar heating of buildings is not limited only to the daylight hours and may extend into the night hours. With the appropriate design of the heating system, solar heating may provide hot water and heat for buildings throughout the year and save a significant amount of primary energy resources.

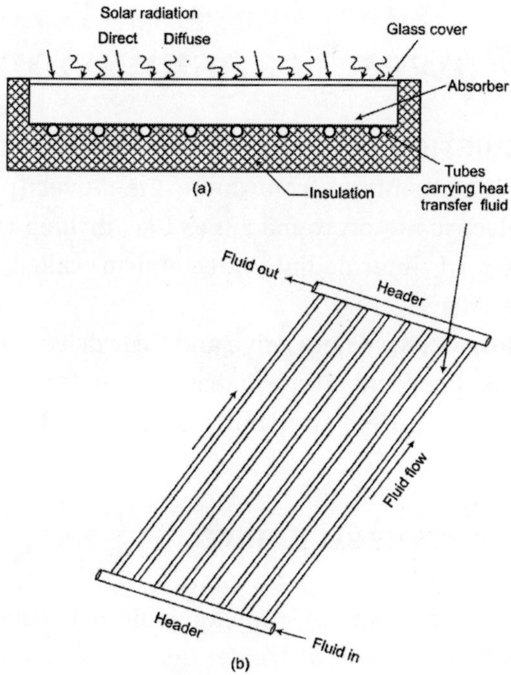

Figure 1.22 Liquid Flat Plate Collector

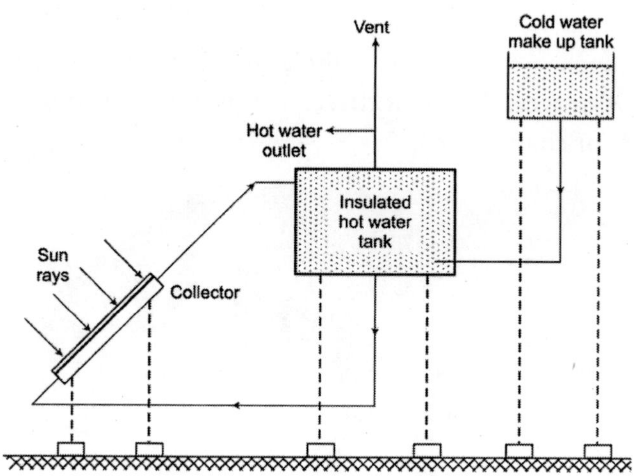

Figure 1.23 Solar Water Heater

Chapter 2

Power Transmission

2.1 INTRODUCTION

The power is transmitted from the prime mover (power source), for example electric motor, to machines like, drilling machine, lathe, etc., by means of intermediate mechanism called drives. Most commonly used drives are:

Belt drive, Rope drive, Chain drive and Gear drive.

The selection of particular type of drive depends on the application, i.e., the amount of power transmitted, distance between the two shafts, etc,

2.2 COMPONENTS OF BELT DRIVE

The belt drives are used to transmit power or motion from one shaft to another by means of a inextensible belt running over two pulleys (Fig. 3.1). A pulley is a circular disc having a hole at center so as to accommodate a shaft in it. Motion is transmitted due to frictional forcebetween the pulleys and the belt surface.Belt drives run smoothly and with little noise, and provide shock absorption for motors, loads, and bearings when the force and power needed changes. A drawback to belt drives is that they transmit less power than gears or chain drives.

Figure 2.1 Belt drives

A simple arrangement of belt drive is shown in figure 3.2. The belt drive consists of two pulleys mounted on two different shafts and a continuous inextensible belt running over these two pulleys. One shaft called the driving shaft receives power from the mains and transmits to another shaft called driven shaft. The pulley mounted on the driving shaft is called driving pulley or driver while the other pulley mounted on a shaft to which the power is to be transmitted is called the driven pulley or follower. The belt passing over the pulley is kept in tension so as to avoid slip over the pulley. This helps in transmitting power effectively from one shaft to another.

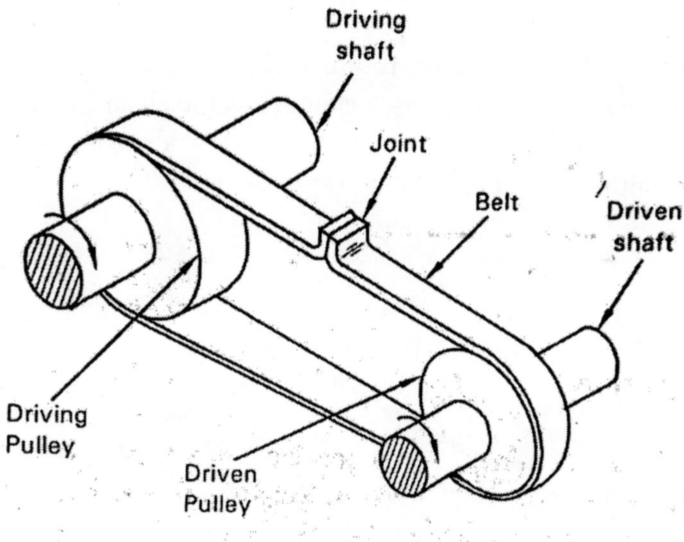

Figure 2.2Belt drive arrangement

2.3 TYPES OF BELT DRIVE

There are two types of belt drives – open belt, and cross belt drives.

In open belt drive, the driving and the driven pulley are rotating in the same direction. When shafts are placed apart, the lower side of the belt should be tight side and upper side must be the slack side. When the upper side becomes slack side, it will sag due to its own weight and thus increases the arc of contact which in turn increases the capacity of the drive.

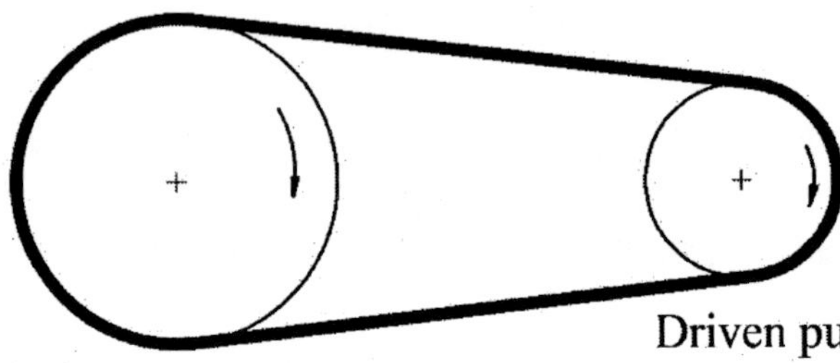

Figure 2.3 Open belt drive

In cross belt drives they rotate in opposite directions. At the junction where the belt crosses, it rubs against itself and wears off. To avoid excessive wear, the shaft must be placed at a maximum distance from each other and operate at low speeds.

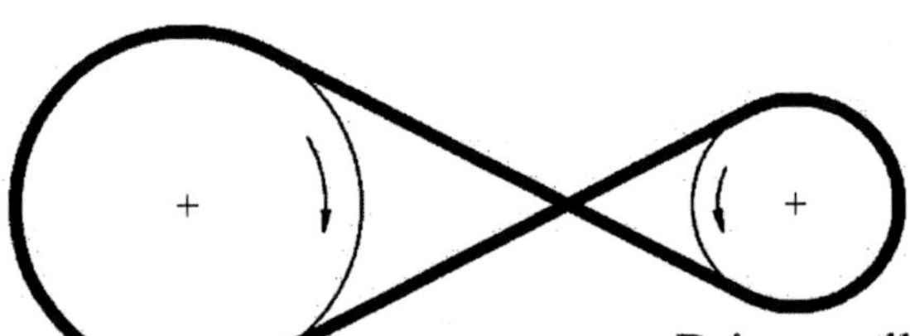

Figure 2.4 Cross belt drive

2.4 VELOCITY RATIO OF BELT DRIVE

The velocity ratio of a belt drive is defined as the ratio of the speed of the driven pulley to the speed of the driving pulley.

When there is no slip, the peripheral velocity of the two pulleys should be same.

Let d_1 and d_2 be the diameter of the driving pulley and driven pulley respectively, let N_1 and N_2 be their speeds in rpm.

Linear speed of the belt = [circumferential speed of the driving pulley] = [circumferential speed of the driven pulley]

$$= [\pi d_1N_1] = [\pi d_2N_2]$$

$$d_1N_1 = d_2N_2$$

$$\text{Velocity ratio} = N2/N1 = d_1/d_2$$

Considering thickness of belt 't',

$$\text{Velocity ratio} = N_2/N_1 = (d_1+ t)/ (d_2+ t)$$

2.5 SLIP IN BELT DRIVE

The difference between the tensions in the tight and slack sides of the belt is equal to the force of friction. When this condition exists in a belt drive, the friction between the pulley and the belt in contact with it will provide the necessary frictional grip to prevent the sliding of the belt over the pulley. Suppose the difference between the tension in the tight and slack sides of the belt is greater than the force of friction, then the belt begins to slide over the surface of the pulley. This sliding of the belt which causes a relative motion between the pulley and belt is called slip.

2.6 CREEP IN BELT DRIVE

Any material when subjected to tension elongates, and so is the belt material. This elongation is directly proportional to the tension in the belt. Tensions on the two sides of the belt passing over the pulley are different, on the tight side the tension is more, and on the slack side, the tension is less. Hence, there is a gradual increase in the length of belt over the surface of the pulley. This stretching results in relative motion between the belt and pulley surfaces. This phenomenon is called creep. Any material when subjected to tension elongates, and so is the belt material. This elongation is directly proportional to the tension in the belt. Tensions on the two sides of the belt passing over the pulley are different, on the tight side the tension is more, and on the slack side, the tension is less. Hence, there is a gradual increase in the length of belt over the surface of the pulley. This stretching results in relative motion between the belt and pulley surfaces. This phenomenon is called creep.

2.7 LENGTH OF BELT FOR OPEN BELT DRIVE

An open belt drive is shown in figure 3.5.

Let

r_1 = radius of larger pulley

r_2 = radius of larger pulley

X = center distance between two pulleys

L = length of belt

Angle of lap of the smaller pulley = $\pi - 2\Phi$ radians

Angle of lap of the larger pulley = $\pi + 2\Phi$ radians

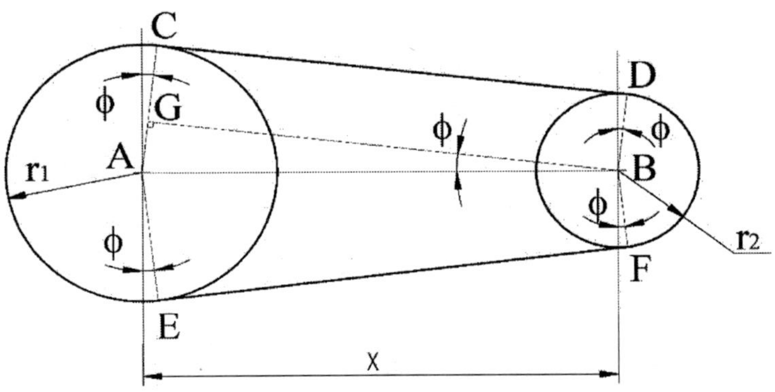

Figure 2.5 Length of belt for open belt drive

Length of belt $L = \text{Arc } EC + CD + \text{Arc } DF + EF$

$$= (\pi + 2\Phi) r_1 + 2\,BG + (\pi - 2\Phi) r_2$$

$$= \pi (r_1 + r_2) + 2\Phi (r_1 - r_2) + 2X \cos\Phi \qquad CD=EF=BG$$

From geometry, $\sin\Phi = \dfrac{AG}{AB} = \dfrac{(r_1 - r_2)}{X}$

$$\Phi = \sin^{-1}\frac{(r_1 - r_2)}{X} = \frac{(r_1 - r_2)}{X}$$

$$\cos\Phi = (1 - \sin^2\Phi)^{1/2}$$

$$= \left[1 - \frac{1}{2}\sin^2\Phi\right]$$

$$= \left[1 - \frac{1}{2}\left\{\frac{r_1 - r_2}{X}\right\}^2\right]$$

Substituting Φ andcos Φ in equation, we get

$$L = \pi(r_1 + r_2) + \frac{2(r_1 - r_2)^2}{X} + 2X\left[1 - \frac{1}{2}\left\{\frac{r_1 - r_2}{X}\right\}^2\right]$$

$$L = \pi(r_1 + r_2) + \frac{2(r_1 - r_2)^2}{X} + 2X - \frac{(r_1 - r_2)^2}{X}$$

$$L = \pi(r_1 + r_2) + \frac{(r_1 - r_2)^2}{X} + 2X$$

2.8 LENGTH OF BELT FOR CROSSED BELT DRIVE

A crossed belt drive is shown in figure 3.6.

Let

r_1 = radius of larger pulley

r_2 = radius of larger pulley

X = center distance between two pulleys

L = length of belt

Angle of lap of the smaller pulley = π + 2Φ radians

Angle of lap of the larger pulley = π + 2 Φ radians

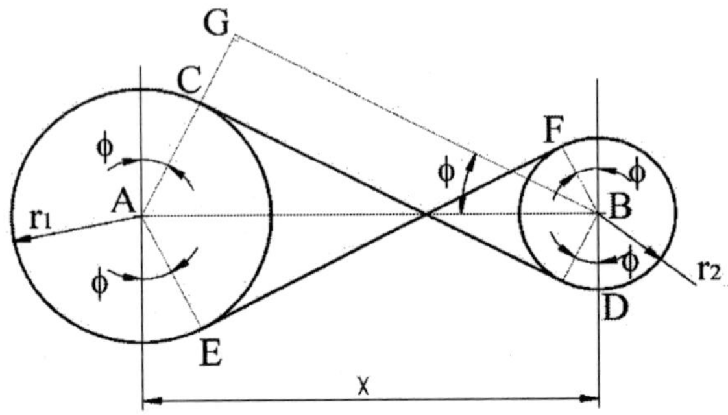

Figure 2.6 Length of belt for cross belt drive

Length of belt $L = \text{Arc } EC + CD + \text{Arc } DF + FE$

$$= (\pi + 2\Phi) r_1 + 2\,BG + (\pi + 2\Phi) r_2$$

$$= \pi (r_1 + r_2) + 2\Phi (r_1 + r_2) + 2X \cos \Phi$$

From geometry, $\sin \Phi = \frac{AC}{AB} = \frac{(r_1 + r_2)}{X}$

$$\Phi = \sin^{-1} \frac{(r_1 + r_2)}{X} = \frac{(r_1 + r_2)}{X}$$

$$\cos \Phi = (1 - \sin^2 \Phi)^{1/2}$$

$$= \left[1 - \frac{1}{2} \sin^2 \Phi\right]$$

$$= \left[1 - \frac{1}{2}\left\{\frac{r_1 + r_2}{X}\right\}^2\right]$$

$$L = \pi(r_1 + r_2) + \frac{2(r_1 + r_2)^2}{X} + 2X\left[1 - \frac{1}{2}\left\{\frac{r_1 + r_2}{X}\right\}^2\right]$$

$$L = \pi(r_1 + r_2) + \frac{2(r_1 + r_2)^2}{X} + 2X - \frac{(r_1 + r_2)^2}{X}$$

$$L = \pi(r_1 + r_2) + \frac{(r_1 + r_2)^2}{X} + 2X$$

2.9 INITIAL TENSION BELT DRIVE

When a belt is wound round the two pulleys (i.e., driver and follower), its two ends are joined together; so that the belt may continuously move over the pulleys, since the motion of the belt from the driver and the follower is governed by a firm grip, due to friction between the belt and the pulleys. In order to increase this grip, the belt is tightened up. At this stage, even when the pulleys are stationary, the belt is subjected to some tension, called initial tension.

When the driver starts rotating, it pulls the belt from one side (increasing tension in the belt on this side) and delivers it to the other side (decreasing the tension in the belt on that side). The increased tension in one side of the belt is called tension in tight side and the decreased tension in the other side of the belt is called tension in the slack side.

Let T_0 = Initial tension in the belt,

T_1 = Tension in the tight side of the belt,

T_2 = Tension in the slack side of the belt,

Initial tension in the belt is given by

$$T_0 = \frac{T_1 + T_2}{2}$$

2.10 BELT TENTIONS IN FLAT BELT DRIVE

Consider flat belt wound around a driven pulley as shown in figure 3.7 (a). let the driven pulley rotates in clockwise direction.

Let T_1 = tension on tight side of the belt

T_2 = tension on slack side of the belt

θ = arc of contact between belt and pulley

Consider a small element PQ of the belt. Let $\delta\theta$ be the angle subtended by the element PQ. The element PQ is in the equilibrium under the action of the following forces. Refer figure 3.7 (b),

Slack side tension (T) acting at P

tight side tension (T + δT) acting at Q

Normal reaction (R) exerted by the pulley on belt element PQ

Friction force (F = μR) acting perpendicular to the normal reaction (R),

Resolving all forces horizontally, $T\sin\frac{\delta\theta}{2}+(T+\delta T)\sin\frac{\delta\theta}{2}=R$

$$T\sin\frac{\delta\theta}{2}+T\sin\frac{\delta\theta}{2}+\delta T\sin\frac{\delta\theta}{2}=R \quad \text{----[1]}$$

Since the angle $\frac{\delta\theta}{2}$ is very small, $\sin\frac{\delta\theta}{2}=\frac{\delta\theta}{2}$

$\therefore$ equation (1) becomes, $T\frac{\delta\theta}{2}+T\frac{\delta\theta}{2}+\delta T\frac{\delta\theta}{2}=R$

$$R=2T\frac{\delta\theta}{2}+\delta T\frac{\delta\theta}{2}$$

Neglecting $\delta T\cdot\frac{\delta\theta}{2}$ for small angles, we get $R = T\delta\theta$ ----[2]

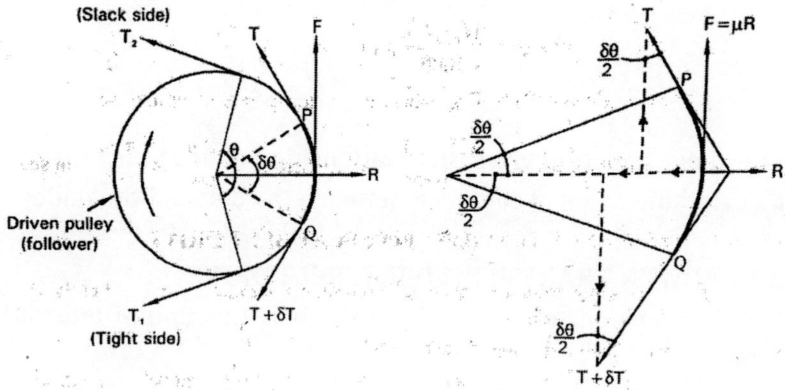

(a) Flat belt wound around driven pulley (b) Forces acting on belt element PQ

Figure 2.7Belt tensions

Resolving all forces vertically, $T\cos\frac{\delta\theta}{2} + \mu R = (T+\delta T)\cos\frac{\delta\theta}{2}$

$$T\cos\frac{\delta\theta}{2} + \mu R = T\cos\frac{\delta\theta}{2} + \delta T\cos\frac{\delta\theta}{2}$$

$$\mu R = \delta T\cos\frac{\delta\theta}{2} \quad \text{- - - - [3]}$$

since $\frac{\delta\theta}{2}$ is small, $\cos\frac{\delta\theta}{2} = 1$

$\therefore$ Equation (3) becomes, $\mu R = \delta T$

$$\text{or } R = \frac{\delta T}{\mu}$$

Equating (2) and (4), we have, $T\delta\theta = \frac{\delta T}{\mu}$

$$\text{or } \frac{\delta T}{T} = \mu\delta\theta$$

Integrating between limits T_2 and T_1 and from 0 to θ respectively, we have,

$$\int_{T_2}^{T_1}\frac{\delta T}{T} = \mu\int_0^{\theta}\delta\theta$$

$$\log_e\left(\frac{T_1}{T_2}\right) = \mu\theta$$

$$\text{or} \quad \frac{T_1}{T_2} = e^{\mu\theta}$$

The above equation gives the relationship between the tensions T_1 andT_2. And the angle of contact θ between the belt and the pulley.

2.11 POWER TRANSMITTED IN A BELT DRIVE

Let T_1 = tension on tight side of the belt (maximum tension) in Newton (*N*)

T_2 = tension on slack side of the belt in *N*

d_1 = diameter of driver pulley in *m*

d_2 = diameter of driven or follower pulley in *m*

The effective turning force acting on the circumference of the follower is the difference in tensions on the tight side (T_1) and slack side (T_2) of the belt.

∴ Driving force is given by $F = (T_1 - T_2)$ Newton.

$$\text{or} \quad F = \frac{(T_1 - T_2)}{1000}\ kN \qquad \text{---- [1]}$$

we know that, Power (P) = F v, where v = velocity of the belt in m/sec.

$$\therefore \text{ equation (1) becomes } P = \frac{(T_1 - T_2)\cdot v}{1000}\ \text{kW, where } v = \frac{\pi d_1 n_1}{60} \text{ or } \frac{\pi d_2 n_2}{60}\ \text{m/sec}$$

2.12 APPLICATIONS OF BELT DRIVE

Belt drives are having wide range of applications of power transmission starting from industrial, automobile, marine, aerospace, farming sector etc.

Chapter -3

Gears and Gera Drives

3.1 GEAR DRIVE

A gear is a toothed wheel that engages another toothed mechanism to change speed or the direction of transmitted motion. Combination of gears, known as gear train is generally used for one of four different reasons:

1. To increase or decrease the speed of rotation,
2. To change the amount of force or torque,
3. To move rotational motion to a different axis (i.e. parallel, right angles, rotating, linear etc.), and
4. To reverse the direction of rotation.

Figure 3.1 Gear trains inside a mechanical wrist watch

Gears are compact, positive-engagement, power transmission elements capable of changing the amount of force or torque. Sports cars go fast (have speed) but cannot pull any weight. Big trucks can pull heavy loads (have power) but cannot go fast. Gears cause this. Gears are generally selected and manufactured using standards established by American Gear Manufacturers Association (AGMA) and American National Standards Institute (ANSI).

A gear train is a mechanical system formed by mounting gears on a frame, so the teeth of the gears engage. Gear teeth are designed to ensure the pitch circles of engaging gears roll on each other without slipping, providing a smooth transmission of rotation from one gear to the next.

The main purpose of a gear train is to increase torque or speed. The arrangement of the driver and driven gears determine if the gear train will increase torque or speed. To increase output torque using a gear train, a power source should be directly connected to a smaller gear and used to drive a larger gear.

3.2 GEAR TYPES, CHARACTERISTICS & APPLICATIONS

The gears can be classified according to:

1. the position of shaft axes
2. the peripheral velocity
3. the type of gears
4. the teeth position

According to the position of shaft axes:

Gears may be classified according to the relative position of the axes of revolution. Theaxes may be:

1. ***Parallel*** shafts where the angle between driving and driven shaft is 0 degree.Examples include spur gears, single and double helical gears.

2. ***Intersecting*** shafts where there is some angle between driving and driven shaft.Examples include bevel and miter gear.

3. ***Non-intersecting and non-parallel*** shafts where the shafts are not coplanar.Examples include the hypoid and worm gear.

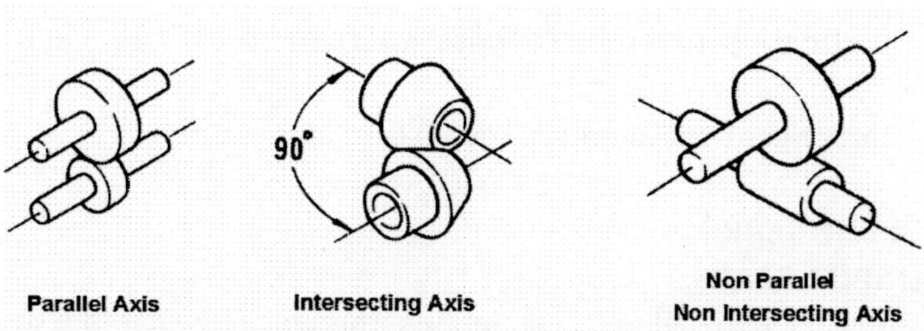

Figure 3.2 Classification of gears according to position of shafts

. According to peripheral velocity:

Gears can be classified as:

1. **Low** velocity type, if their peripheral velocity lies in the range of 1 to 3 m/sec.

2. **Medium** velocity type, if their peripheral velocity lies in the range of 3 to 15m/sec.

3. **High** velocity type, if their peripheral velocity exceeds 15 m/sec.

According to type of gears:

Gears can be classified as external gears, internal gears, and rack and pinion.

1. **External gears** mesh externally - the bigger one is called "gear" and the smallerone is called "pinion".

2. **Internal gears** mesh internally - the larger one is called "annular" gear and thesmaller one is called "pinion".

3. **Rack and pinion type –** converts rotary to linear motion or vice versa. There is astraight line gear called "rack" on which a small rotary gear called "pinion" moves.

According to teeth position:

Gears are classified as straight, inclined and curved.

1. **Straight gear teeth** are those where the teeth axis is parallel to

the shaft axis.

2. Inclined gear teeth are those where the teeth axis is at some angle.

3. Curve gear teeth are curved on the rim's surface.

TYPEOF GEARS

(a) SPUR GEARS

Spur gears are used to transmit power between two parallel shafts. The teeth on these gears are cut straight and are parallel to the shafts to which they are attached.

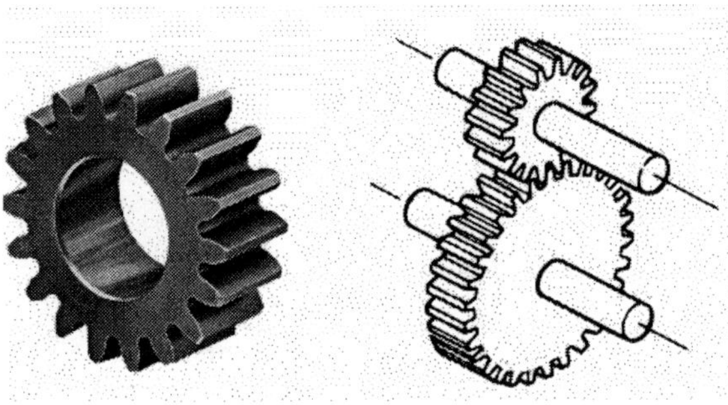

Figure 3.3 Spur gears

Characteristics:

- Simplest and most economical type of gear to manufacture.
- Speed ratios of up to 8 (in extreme cases up to 20) for one step (single reduction)design; up to 45 for two step design; and up to 200 for three-step design.

Limitations:

- Not suitable when a direction change between the two shafts is required.
- Produce noise because the contact occurs over the full face width of the matingteeth instantaneously.

(b) HELICAL GEARS

Helical gears resemble spur gears, but the teeth are cut at an angle rather than parallel to the shaft axis like on spur gears. The angle that the helical gear tooth is on is referred to as the helix angle. The angle of helix depends upon the condition of the shaft design and relative position of the shafts. To ensure that the gears run smoothly, the helix angleshould be such that one end of the gear tooth remains in contact until the opposite endof the following gear tooth has found a contact. For parallel shafts, the helix angle shouldnot exceed 20 degrees to avoid excessive end thrust.

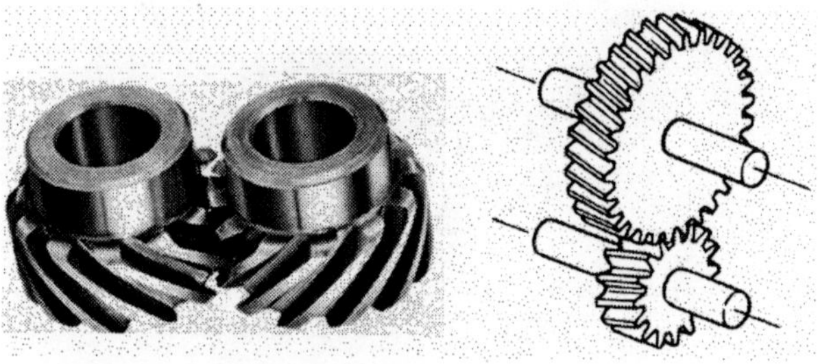

Figure 3.4 Helical gears

Characteristics:

The longer teeth cause helical gears to have the following differences from spur gears ofthe same size:

- Tooth strength is greater because the teeth are longer than the teeth of spur gearof equivalent pitch diameter.
- Can carry higher loads than can spur gears because of greater surface contacton the teeth.
- Can be used to connect parallel shafts as well as non-parallel, non-intersectingshafts.
- Quieter even at higher speed and are durable.

Limitations:

- Gears in mesh produce thrust forces in the axial directions.

- Expensive compared to spur gears.

(c) BEVEL GEARS

A bevel gear is shaped like a section of a cone and primarily used to transfer power between intersecting shafts at right angles. The teeth of a bevel gear may be straight or spiral. Straight gear is preferred for peripheral speeds up to 1000 feet per minute; above that they tend to be noisy.

Figure 3.5 Bevel gears

Characteristics:

- Designed for the efficient transmission of power and motion between intersectingshafts. A good example of bevel gears is seen as the main mechanism for ahand drill. As the handle of the drill is turned in a vertical direction, the bevelgears change the rotation of the chuck to a horizontal rotation.
- Permit a minor adjustment during assembly and allow for some displacementdue to deflection under operating loads without concentrating the load on the endof the tooth.

(d) MITTER GEARS

Mitter gears are identical to bevel gears with the exception that both gears always havethe same number of teeth.

Figure 3.6 Mitter gears

Characteristics:

- They provide a steady ratio; other characteristics are similar to bevel gears.
- They are used as important parts of conveyors, elevators and kilns.

Limitations

- Gear ration is always 1 to 1 and therefore not used when an application calls fora change of speed.

(e) HYPOID GEARS

Hypoid gears are a modification of the spiral bevel gear with the axis offset. Thedistinguishing feature of hypoid gears is that the shafts of the pinion and ring gear may continue past each other, never having their axis intersecting.

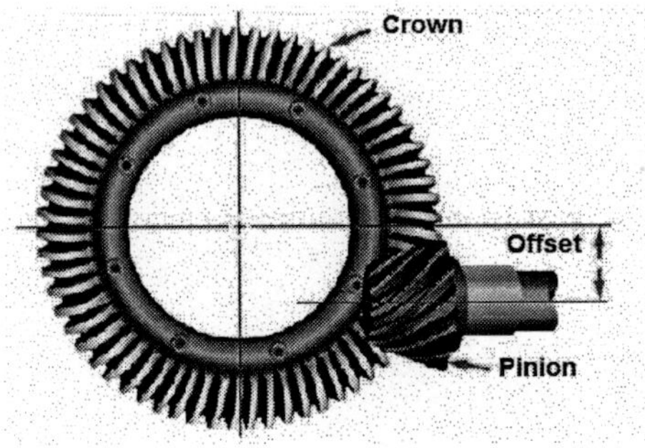

Figure 3.7 Hypoid gears

The major advantages of the hypoid gear design are that the pinion diameter is increased, and it is stronger than a corresponding bevel gear pinion. The increased diameter size of the pinion permits the use of comparatively high gear ratios and is extremely useful for non-intersecting shaft requirements such as automotive applications where the offset permits lowering of the drive shaft.

(f) WORM GEARS

Worm gears are used to transmit power between two shafts that are at right angles to each other and are non-intersecting.

Worm gears are special gears that resemble screws, and can be used to drive spur gears or helical gears. Worm gearing is essentially a special form of helical gearing in which the teeth have line contact and the axes of the driving and driven shafts areusually at right angles and do not intersect.

Figure 3.8 Bevel gears

Characteristics:

• Meshes are self-locking. Worm gears have an interesting feature that no other gear set has: the worm can easily turn the gear, but the gear cannot turn the worm. This is because the angle on the worm is so shallow that when the gear tries to spin it, the friction between the gear and the worm holds the worm in place. This feature is useful for machines such as conveyor systems, in whichthe locking feature can act as a brake for the conveyor when the motor is not turning.

• Worm gear is always used as the input gear, i.e. the torque is applied to the inputend of the worm shaft by a driven sprocket or electric motor.

• Best suited for applications where a great ratio reduction is required between the driving and driven shafts. It is common for worm gears to have reductions of 20:1, and even up to 300:1 or greater.

Limitations:

• Yield low efficiency because of high sliding velocities across the teeth, thereby causing high friction losses.

• When used in high torque applications, the friction causes the wear on the gear teeth and erosion of the restraining surface.

(g) RACKS (STRAIGHT GEARS)

The rack is a bar with a profile of the gear of infinite diameter, and when used with ameshing pinion, enables the rotary to linear movement or vice versa.

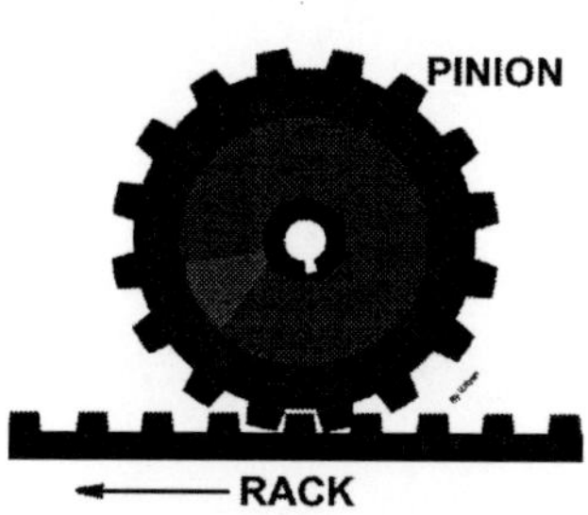

Figure 3.9 Rack and pinion

Characteristics:

- Racks with machined ends can be joined together to make any desired length.
- The most well-known application of a rack is the rack and pinion steering system used on many cars in the past. The steering wheel of a car rotates the gear that engages the rack. The rack slides right or left, when the gear turns, depending onthe way we turn the wheel. Windshield wipers in cars are powered by a rack and pinion mechanism.

3.5 GEARS FUNDAMENTALS & TERMINOLOGY

In this section, we will discuss the gear fundamentals, considering spur gears. It is the most common and the simplest form, and hence the most comprehensible. The same principles apply to spiral gears and bevel gears too.

A gear can be defined in terms of its pitch, pressure angle and number of teeth. Let's discuss few terms here:

Pitch Circle Diameter (d) - This is the diameter of a circle about which the gear tooth geometry is designed or constructed. The pitch circle is the imaginary circle found at the point where the teeth of two gears mesh. The diameter of the pitch circle is called the pitch diameter.

Outside Diameter (OD) - The outside circle is the distance around the outer edge of thegear's teeth. The diameter of the outside circle is called the outside diameter.

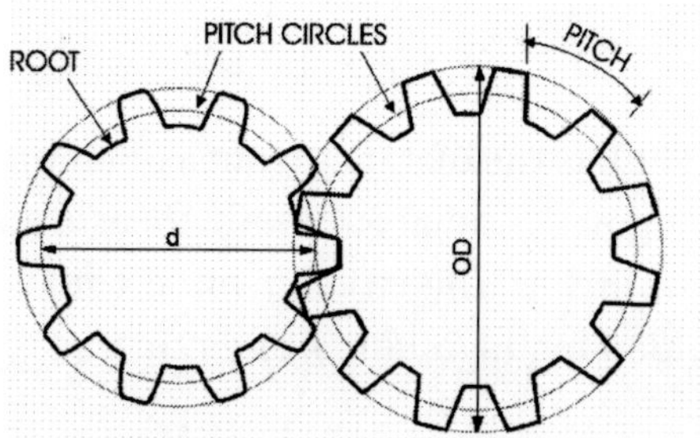

Figure 3.10 Terminology of gears in mesh

Root - The root is the bottom part of a gear wheel.

Pitch - Pitch is a measure of tooth spacing along the pitch circle. It is the distance between any point on one tooth and same point on the next tooth. It is expressed in the following forms:

Diametral Pitch (P_d) is the number of teeth per inch of the pitch diameter and is also an index of tooth size. It is given as:

Where:

$$P_d = \frac{Z}{d}$$

P_d = diametral pitch
Z = number of teeth
d = pitch circle diameter in inches

A large diametral pitch indicates a small tooth and vice versa. Another way of sayingthis; larger gears have fewer teeth per inch of diametral pitch.

Module (m) is the metric equivalent of diametral pitch, i.e. the pitch diameter (in mm) divided by the number of teeth, but unlike diametral pitch, the higher number, the larger the teeth. Meshing gears must have the same module:

$$m = \frac{1}{P_d} = \frac{d}{Z}$$

Circular Pitch (Pc): is the distance from a point on one tooth to the corresponding poingon the adjacent tooth, measured along the pitch circle. Calculated in inches, the circular pitch equals the pitch circle circumference divided by the number of teeth:

$$P_c = \frac{\text{Circumfrence } (\pi d)}{\text{Number of teeth } (Z)}$$

Because the circular pitch is directly proportional to the module and inversely proportional to the diametral pitch, meshing teeth must have the same circular pitch.

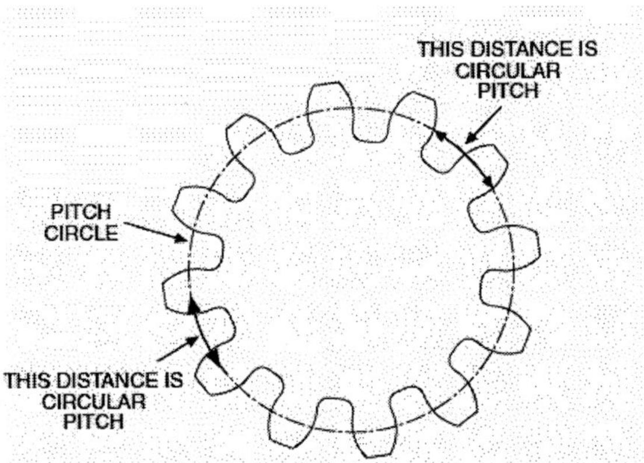

Figure 3.12 Gear terminology

Relationship between Circular Pitch and Diametral Pitch:

$$\text{From } P_c = \frac{\pi d}{Z} \quad \text{and} \quad P_d = \frac{Z}{d}$$

We have,

$$P_d P_c = \pi$$

The product of the circular pitch and the diametral pitch is equal to **pi** (π).

Number of Teeth (N): The number of gear teeth is related to the diametral pitch and the pitch circle diameter by equation,

$Z = d \times P_d$.

Tooth Size: Diametral pitch, module and circular pitch are all indications of tooth size; ratios which determine the number of teeth in a gear for a given pitch diameter. In designing a gear set, the number of teeth in each member is of necessity. *As a rule of thumb, teeth should be large and low in number for heavily loaded gears and small and numerous for smooth operation.*

Center Distance (CD)

Center Distance is the distance between the centers of the shaft of one spur gear to the center of the shaft of the other spur gear. The standard center distance between twospur gears is one-half the sum of their pitch diameters.

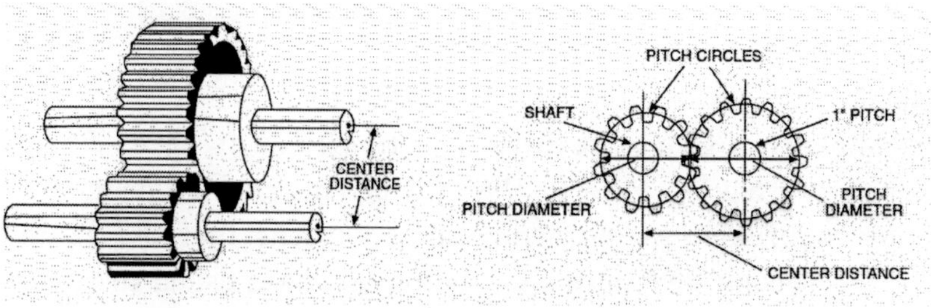

Figure 3.13 Gears in mesh3.3 GEAR TRAIN

A gear train is a combination of gears used to transmit power from one shaft to another. The gear to which the force is first applied is called the driver and the final gear on the train to which the force is transmitted is called the driven gear. Any gears between the driver and the driven gears are called the idlers. Conventionally, the smaller gear is the **Pinion** (B) and the larger one is the **Gear** (A). In most applications, the pinion is the driver; this reduces speed but increases torque.

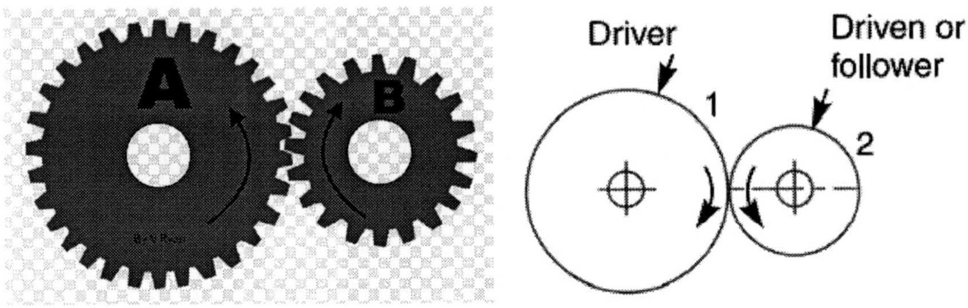

Figure 3.14 Gear train

When gears are connected together, it is common to talk about the corresponding gear ratio. Gear ratio is the ratio of the number of teeth on the first gear to the number of teeth on the second. It is the reciprocal of speed ratio, which is the ratio of second gear to first gear. Assuming T_1, T_2 as number of teeth on first (driver) gear and second (driven) gear and N_1, N_2 as rotational speed (in rpm) of the two gears respectively,

Speed ratio = $N_2/N_1 = T_1/T_2$

Gear ratio Tarin value = $N_1/N_2 = T_2/T_1$

Speed ratio = = Speed of driven No. of teeth on driver

If you turn a gear with 6 teeth 3 times and is meshed with a second gear having 18teeth, than the driving gear 18 teeth (6 x 3) will move through the meshed area. This means that the 18 teeth from the second gear also move through the meshed area. If the second gear has 18 teeth, then it only has to rotate once because 18 x 1=18.

Also, the second gear will be turning slower than the first because it is larger, and larger gears turn slower than smaller gears because they have more teeth.

3.4 TYPES OF GEAR TRAINS

1. Simple gear train

2. Compound gear train
3. Planetary gear train

Simple Gear Train

Simple gear trains have only one gear per shaft as shown in Figure 3.16. The simple gear train is used where there is a large distance to be covered between the input shaft and the output shaft.

In a simple gear train,

1. Two external gears of a pair always move in opposite directions.
2. All odd numbered gears move in one direction and all even numbered gears move in the opposite.
3. Speed ratio is negative when the input and the output gears rotate in the opposite directions and it is positive when the two rotate in the same direction.
4. All the gears can be in a straight line or arranged in a zig-zag manner. A simple gear train can also have bevel gears.

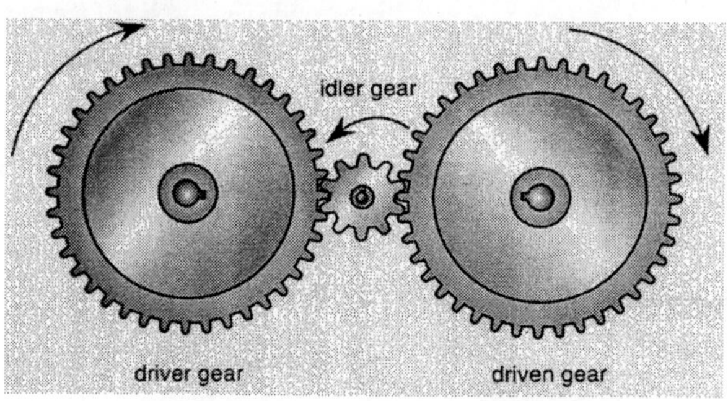

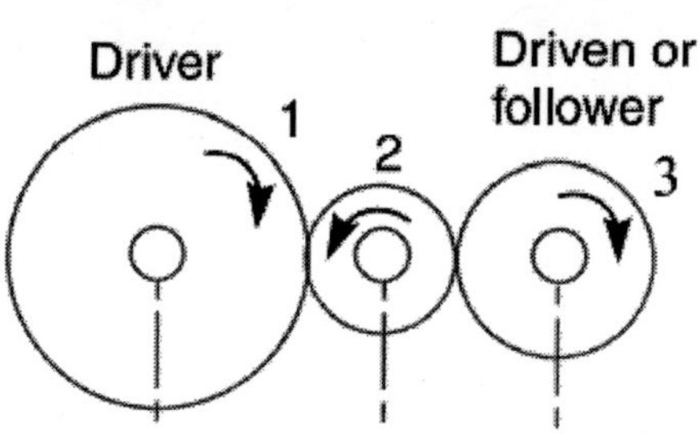

Figure 3.15 Simple gear train

Let T = number of teeth on a gear, N = speed of a gear in rpm

Referring to Figure 3.16, since the driving gear 1 is in mesh with the intermediate gear 2, therefore speed ratio for these two gears is

$$\frac{N_1}{N_2} = \frac{T_2}{T_1} \quad ...(i)$$

Similarly, as the intermediate gear 2 is in mesh with the driven gear 3, therefore speed ratio for these two gears is

$$\frac{N_2}{N_3} = \frac{T_3}{T_2} \quad ...(ii)$$

The speed ratio of the gear train as shown in Figure 3.16 is obtained by multiplying the equations (i) and (ii).

$$\therefore \quad \frac{N_1}{N_2} \times \frac{N_2}{N_3} = \frac{T_2}{T_1} \times \frac{T_3}{T_2} \quad \text{or} \quad \frac{N_1}{N_3} = \frac{T_3}{T_1}$$

i.e.

$$\text{Speed ratio} = \frac{\text{Speed of driver}}{\text{Speed of driven}} = \frac{\text{No. of teeth on driven}}{\text{No. of teeth on driver}}$$

and

$$\text{Train value} = \frac{\text{Speed of driven}}{\text{Speed of driver}} = \frac{\text{No. of teeth on driver}}{\text{No. of teeth on driven}}$$

Thus, it is seen that the intermediate gears have no effect on the speed ratio and, therefore, they are known as idlers.

Compound Gear Train

In a compound gear train at least one of the intermediate shafts in the train must hold two gears as shown in Figure 3.17. Compound gear trains are used when large changes in speed or power output are needed and there is only a small space between the input and output shafts.

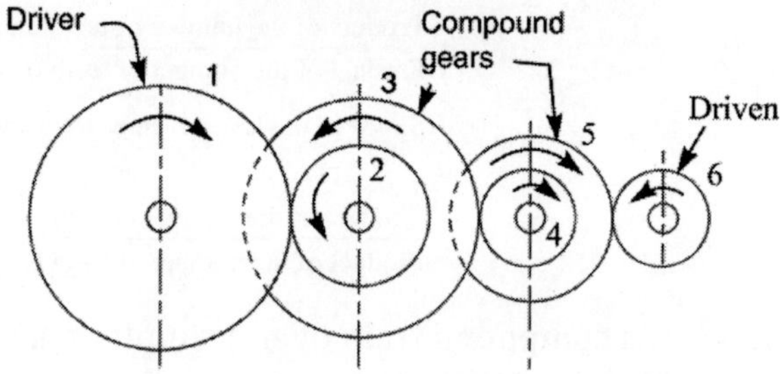

Figure 3.16 Compound gear train

In a compound train of gears, as shown in Figure 3.17, the gear 1 is the driving gear, gears 2 and 3 are compound gears which are mounted on one of the intermediate shaft. The gears 4 and 5 are also compound gears which are mounted on another intermediate shaft and the gear 6 is the driven gear mounted on the driven shaft.

Let T = number of teeth on a gear, N = speed of a gear in rpm

Since gear 1 is in mesh with gear 2, therefore its speed ratio is

$$\frac{N_1}{N_2} = \frac{T_2}{T_1} \qquad ...(i)$$

Similarly, for gears 3 and 4, speed ratio is

$$\frac{N_3}{N_4} = \frac{T_4}{T_3} \qquad ...(ii)$$

and for gears 5 and 6, speed ratio is

The speed ratio of compound gear train is obtained by multiplying the equations (*i*), (*ii*) and (*iii*),

$$\therefore \quad \frac{N_1}{N_2} \times \frac{N_3}{N_4} \times \frac{N_5}{N_6} = \frac{T_2}{T_1} \times \frac{T_4}{T_3} \times \frac{T_6}{T_5} \quad \text{or} \quad {}^{*}\frac{N_1}{N_6} = \frac{T_2 \times T_4 \times T_6}{T_1 \times T_3 \times T_5}$$

Note: Since gears 2 and 3 are mounted on a single shaft, therefore $N_2 = N_3$. Similarly gears 4 and 5 are mounted on a single shaft, therefore N4 = N5.

i.e.

$$\text{Speed ratio} = \frac{\text{Speed of the first driver}}{\text{Speed of the last driven or follower}} = \frac{\text{Product of the number of teeth on the drivens}}{\text{Product of the number of teeth on the drivers}}$$

and

$$\text{Train value} = \frac{\text{Speed of the last driven or follower}}{\text{Speed of the first driver}} = \frac{\text{Product of the number of teeth on the drivers}}{\text{Product of the number of teeth on the drivens}}$$

The advantage of a compound train over a simple gear train is that a much larger speed reduction from the first shaft to the last shaft can be obtained with small gears. If a simple gear train is used to give a large speed reduction, the last gear has to be very large. Usually for a speed reduction in excess of 7 to 1, a simple train is not used and a compound train or worm gearing is employed.

Planetary Gear Train

A planetary transmission system (or Epicyclic gear train) as shown in Figure 3.18, consists normally of a centrally pivoted sun gear, a ring gear and several planet gears which rotate between these. This assembly concept explains the term planetary transmission, as the planet gears rotate around the sun gear as in the astronomical sense the planets rotate around the Sun.

In an epicyclic gear train, the axis of rotation of one or more of the wheels is carried on an arm which is free to revolve about the axis of rotation of one of the other wheels in the train. The diagram shows a Gear B on the end of an arm A. Gear B meshes with Gear Cand revolves around it when the arm is rotated. B is called the planet gear and C the sun.

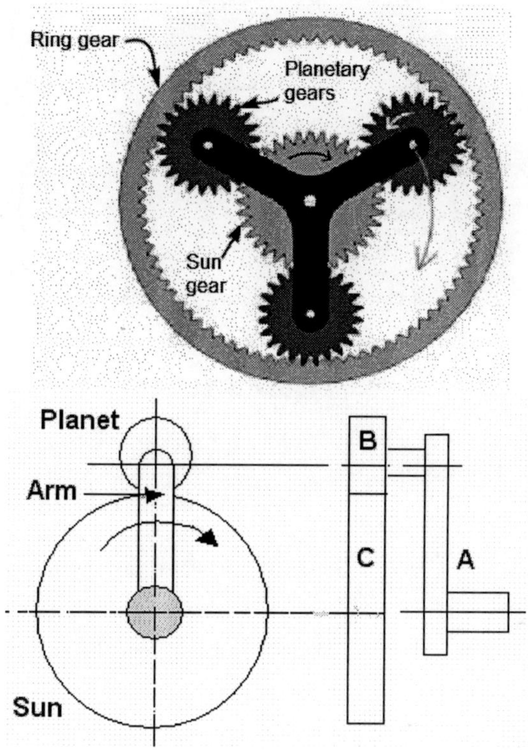

Figure 3.17 Planetary gear train

Planetary gearing or epicyclic gearing provides an efficient means to transfer hightorques utilizing a compact design.

EXAMPLE 3.1

The gearing of a machine tool is shown in Figure 3.19. The motor shaft is connected to gear A and rotates at 975 rpm. The gear wheels B, C, D and E are fixed to parallel shafts rotating together. The final gear F is fixed on the output shaft. What is the speed of gear F?

The number of teeth on each gear are as given below:

Gear:	A	B	C	D	E	F
No. of teeth:	20	50	25	75	26	65

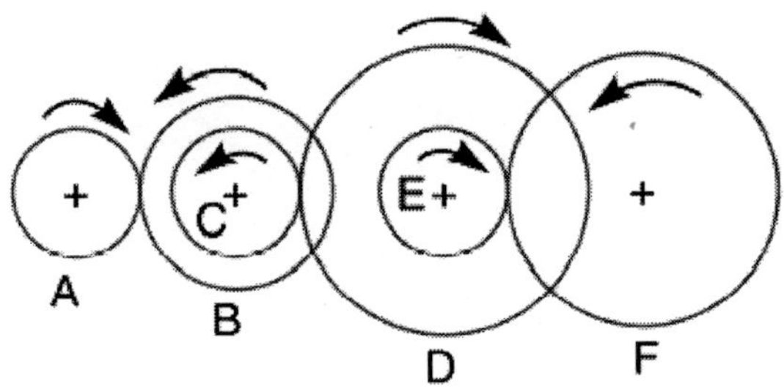

Figure 3.18

Solution:

Given : $N_A = 975$ rpm;

$T_A = 20$; $T_B = 50$; $T_C = 25$; $T_D = 75$; $T_E = 26$; $T_F = 65$

Let N_F = Speed of gear *F*, *i.e.* last driven or follower.

We know that

$$\frac{\text{Speed of the first driver}}{\text{Speed of the last driven}} = \frac{\text{Product of no. of teeth on drivens}}{\text{Product of no. of teeth on drivers}}$$

or
$$\frac{N_A}{N_F} = \frac{T_B \times T_D \times T_F}{T_A \times T_C \times T_E} = \frac{50 \times 75 \times 65}{20 \times 25 \times 26} = 18.75$$

$\therefore$
$$N_F = \frac{N_A}{18.75} = \frac{975}{18.75} = 52 \text{ r. p. m.}$$

EXAMPLE 3.2

In the figure below, Gears B and C represent a compound gear and have the following details:

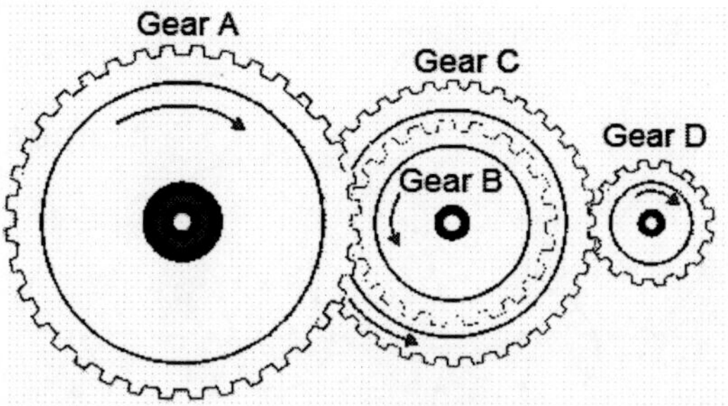

Figure 3.19

Gear A - 120 teeth, Gear B - 40 teeth, Gear C - 80 teeth, Gear D -20 teeth

What is the output in rpm at D, and what is the direction of rotation if Gear A rotates in a clockwise direction at 30 rpm?

Solution:

When answering a question like this, split it into two parts.

Treat Gears A and B as the first part of the question.

Treat Gears C and D as the second part.

Gear ratio AB = driven/driving = 40/120 = 1/3

Gear ratio CD = driven /driving = 20/80 = 1/4

Since the driving Gear A rotates 30 RPM and the Gear B is smaller than Gear A, we canconclude that the rpm for Gear B is 30*3 = 90 rpm

Since Gears B and C represent a compound gear, they have the same rotational speed. Therefore, Gear D speed is obtained by multiplying 4 to Gear C speed of 90 rpm.

Thus, Gear D moves at 90*4 = 360 rpm

OR

1/3 x ¼ = 1/12

Since Gear A moves at 30 rpm and Gear D is smaller, we multiply by 12:30 x 12 = 360 rpm.

EXAMPLE 3.3

Calculate the gear ratio for the compound chain shown below. If the input gear rotates clockwise, in which direction does the output rotate?

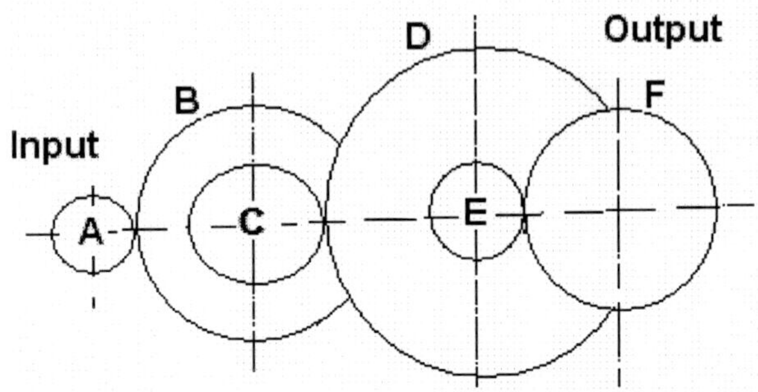

Figure 3.20

Gear A has 20 teeth Gear B has 100 teethGear C has 40 teeth Gear D has 100 teethGear E has 10 teeth Gear F has 100 teeth.

Solution:

The driving teeth are A, C and E

The driven teeth are B, D and F

Gear ratio = (100 x 100 x 100) / (20 x 40 x 10) = 125

Alternatively we can say there are three simple gear trains as follows:First gear GR = 100 / 20 = 5

Second chain GR = 100 / 40 = 2.5

Third chain GR = 100 / 10 = 10

The overall ratio = 5 x 2.5 x 10 = 125

Each chain reverses the direction of rotation so if A is clockwise, B and C rotate anti- clockwise, so D and E rotate clockwise. The output gear F hence rotates anti-clockwise.

More complex compound gear trains can achieve high and low gear ratios in a restrictedspace by coupling large and small gears on the same axle. In this way gear ratios of adjacent gears can be multiplied through the gear train.

3.5 CHAIN DRIVE

Basic features of chain drives include consistent velocity ratio, since no slippage or creep is involved; long life and ability to drive a number of shafts from a single source of power. The chains are made up of number of rigid links which are hinged together by pin joints in order to provide the necessary flexibility for wrapping round the driving and driven wheels. These wheels have projecting teeth of special profile and fit into the corresponding recesses in thelinks of the chain as shown in Figure 3.22.

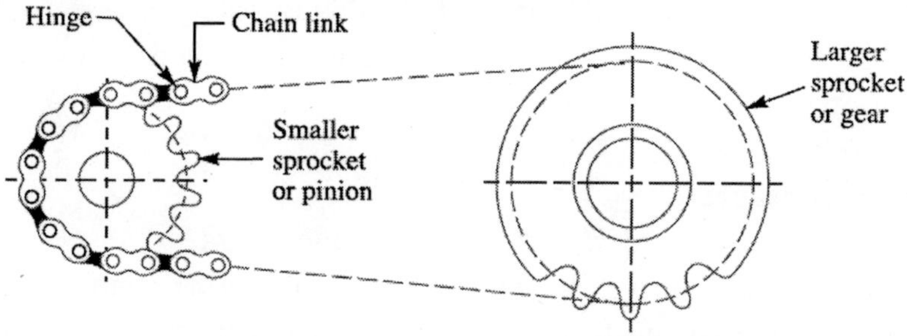

Figure 3.21 Sprocket and chain drive

The toothedwheels are known as sprockets.The sprockets and the chain are thus constrained to move together without slipping and ensures perfect velocity ratio.

The chains are mostly used to transmit motion and power from one shaft to another, when thecentre distance between their shafts is short such as in bicycles, motor cycles, agricultural machinery, conveyors, rolling mills, road rollers etc. The chains may also be used for long centre distance of upto 8 metres. The chains are used for velocities up to 25 m/s and for power upto 110 kW. In some cases,higher power transmission is also possible.

Following are the advantages and disadvantages of chain drive over belt or rope drive:

Advantages:

- As no slip takes place during chain drive, hence perfect velocity ratio is obtained.
- Since the chains are made of metal, therefore they occupy less space in width than a belt or rope drive.
- It may be used for both long as well as short distances.
- It gives a high transmission efficiency (upto 98 percent).
- It gives less load on the shafts.
- It has the ability to transmit motion to several shafts by one chain only.
- It transmits more power than belts.
- It permits high speed ratio of 8 to 10 in one step.
- It can be operated under adverse temperature and atmospheric conditions.

Disadvantages:

- The production cost of chains is relatively high.
- The chain drive needs accurate mounting and careful maintenance, particularly lubricationand slack adjustment.
- The chain drive has velocity fluctuations especially when unduly stretched.

TERMS USED IN CHAIN DRIVE

The following terms are frequently used in chain drive.

Pitch of chain: It is the distance between the hinge centre of a link and the corresponding hinge centre of the adjacent link, as shown in Figure 3.23. It is usually denoted by p.

Pitch circle diameter of chain sprocket:

It is the diameter of the circle on which the hinge centres of the chain lie, when the chain is wrapped round a sprocket as shown in Figure2.2. The points A, B, C, and D are the hinge centres of the chain and the circle drawn through these centres is called pitch circle and its diameter (D) is known as pitch circle diameter.

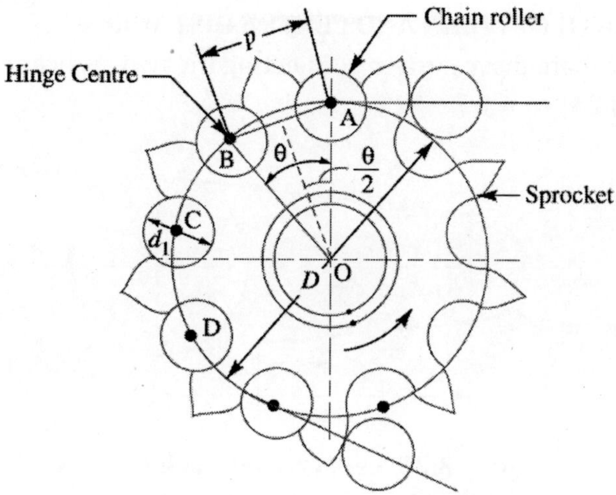

Figure 3.22 Terms used in chain drive

VELOCITY RATIO OF CHAIN DRIVES

The velocity ratio of chain drive is given by

$$V.R. = \frac{N_1}{N_2} = \frac{T_2}{T_1}$$

Where

N_1 = Speed of rotation of smaller sprocket in rpm,

N_2 = Speed of rotation of larger sprocket in rpm,

T_1 = Number of teeth on the smaller sprocket, and

T_2 = Number of teeth on the larger sprocket.

The average velocity of the chain is given by

$$v = \frac{\pi D N}{60} = \frac{T p N}{60}$$

Where

D = Pitch circle diameter of the sprocket in metres, and

p = Pitch of the chain in metres.

3.9 LENGTH OF CHAIN AND CENTRE DISTANCE

An open chain drive system connecting the two sprockets is shown in Figure 3.24

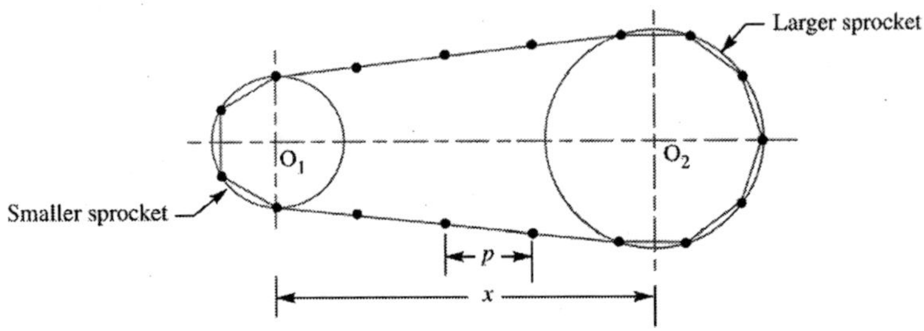

Figure 3.23 Terms used in chain drive

Let

T_1 = Number of teeth on the smaller sprocket, T_2 = Number of teeth on the larger sprocket,

p = Pitch of the chain, and

x = Centre distance.

The length of the chain (L) must be equal to the product of the number of chain links (K) and the pitch of the chain (p). Mathematically,

$L = K.p$

The number of chain links may be obtained from the following expression, *i.e.*

$$K = \frac{T_1 + T_2}{2} + \frac{2x}{p} + \left[\frac{T_2 - T_1}{2\pi}\right]^2 \frac{p}{x}$$

The value of K as obtained from the above expression must be approximated to the nearest even number.

The centre distance is given by

$$x = \frac{p}{4}\left[K - \frac{T_1 + T_2}{2} + \sqrt{\left(K - \frac{T_1 + T_2}{2}\right)^2 - 8\left(\frac{T_2 - T_1}{2\pi}\right)^2}\right]$$

In order to accommodate initial sag in the chain, the value of the centre distance obtained fromthe above equation should be decreased by 2 to 5 mm. For best results, the minimum centre distance should be 30 to 50 times the pitch. The minimum centre distance is selected depending upon the velocity ratio so that the arc of contact of the chain on the smaller sprocket is not less than 120º. It may be noted that larger angle of arc of contact ensures a more uniform distribution of load on the sprocket teeth and better conditions of engagement.

3.6 BALL SCREW

A **ball screw** is a mechanical linear actuator that translates rotational motion to linear motion with little friction. A threaded shaft provides a helical raceway for ball bearings which act as a precision screw. As well as being able to apply or withstand high thrust loads, they can do so with minimum internal friction. They are made to close tolerances and are therefore suitable for use in situations in which high precision is necessary. The ball assembly acts as the nut while the threaded shaft is the screw. In contrast to conventional lead screws, ball screws tend to be rather bulky, due to the need to have a mechanism to re-circulate the balls.

Another form of linear actuator based on a rotating rod is the thread less ball screw, also known as "rolling ring drive". In this design, three (or more) rolling-ring bearings are arranged symmetrically in a housing surrounding a smooth (thread-less) actuator rod or shaft. The bearings are set at an angle to the rod, and this angle determines the direction and rate of linear motion per revolution of the rod. An advantage of this design over the conventional ball screw or lead screw is the practical elimination of backlash and loading caused by pre load nuts.

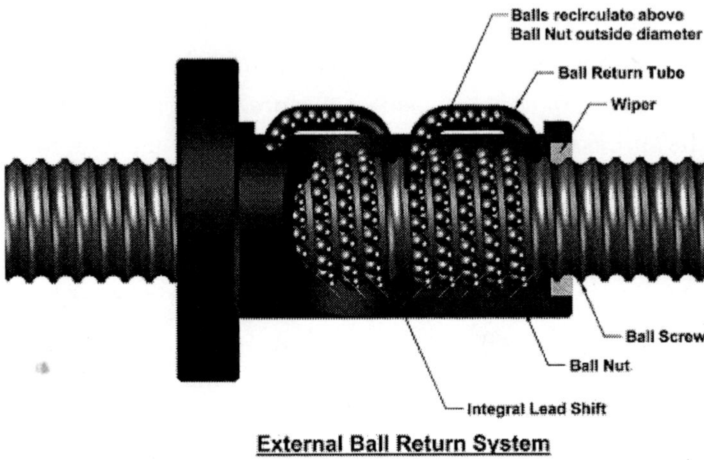

Figure 3.24 Ball Screw

BALL RETURN SYSTEMS

The circulating balls travel inside the thread form of the screw and nut, and balls are recirculated through various types of return mechanisms. If the ball nut did not have a return mechanism the balls would fall out of the end of the ball nut when they reached the end of the nut. For this reason several different re-circulation methods have been developed.

An external ball nut employs a stamped tube which picks up balls from the raceway by use of a small pick up finger. Balls travel inside of the tube and are then replaced back in the thread raceway.

An internal button ball nut employs a machined or cast button style return which allows balls to exit the raceway track and move one thread then re-enter the raceway.

An end cap return ball nut employs a cap on the end of the ball nut. The cap is machined to pick up balls out of the end of the nut and direct them down holes which are bored transversely down the ball nut. The complement cap on the other side of the nut directs balls back into the raceway.

The returning balls are not under significant mechanical load and the return path may incorporate injection moulded low-friction plastic parts.

THREAD PROFILE

To obtain proper rolling action of the balls, as in a standard ball bearing, it is necessary that, when loaded in one direction, the ball makes contact at one point with the nut, and one point with the screw. In practice, most ball screws are designed to be lightly pre-loaded, so that there is at least a slight load on the ball at four points, two in contact with the nut and two in contact with the screw. This is accomplished by using a thread profile that has a slightly larger radius than the ball, the difference in radii being kept small (e.g. a simple V thread with flat faces is unsuitable) so that elastic deformation around the point of contact allows a small, but non-zero contact area to be obtained, like any other rolling element bearing. To this end, the threads are usually machined as a "gothic arch" profile. If a simple semicircular thread profile were used, contact would only be at two points, on the outer and inner edges, which would not resist axial loading.

APPLICATIONS

Ball screws are used in aircraft and missiles to move control surfaces, especially for electric fly by wire, and in automobile power steering to translate rotary motion from an electric motor to axial motion of the steering rack. They are also used in machine tools, robots and precision assembly equipment. High precision ball screws are used in steppers for semiconductor manufacturing.

ADVANTAGES

Low friction in ball screws yields high mechanical efficiency compared to alternatives. A typical ball screw may be 90 percent efficient, versus 20 to 25 percent efficiency of an Acme lead screw of equal size. Lack of sliding friction between the nut and screw lends itself to extended lifespan of the screw assembly (especially in no-backlash systems), reducing downtime for maintenance and parts replacement, while also decreasing demand for lubrication. This, combined with their overall performance benefits and reduced power requirements, may offset the initial costs of using ball screws.

Ball screws may also reduce or eliminate backlash common in lead screw and nut combinations. The balls may be pre-loaded so that there is no "wiggle" between the ball screw and ball nut. This is particularly desirable in applications where the load on the screw varies quickly, such as machine tools.

DISADVANTAGES

Depending upon their lead angle, ball screws can be back-driven due to their low internal friction (i.e., the screw shaft can be driven linearly to rotate the ball nut). They are usually undesirable for hand-fed machine tools, as the stiffness of a servo motor is required to keep the cutter from grabbing the work and self-feeding, that is, where the cutter and work piece exceed the optimum feed rate and effectively jam or crash together, ruining the cutter and work piece. Cost is also a major factor as Acme screws are cheaper to manufacture

Chapter 4

Machine Tools and Robotics

4.1 Introduction

Machining is a process in which a piece of raw material is cut into a desired shape and size using sharp cutting tools. It is a process in which a material is cut to a desired final shape and size by a controlled material-removal process.

Machining is a manufacturing term encompassing a broad range of technologies and techniques. It can be roughly defined as the process of removing material from a workpiece using power-driven machine tools to shape it into an intended design

Figure 4.1: Machining process (Milling operation)

History of Machining Process

Machining is considered as one of the most important manufacturing processes. The industrial revolution and the boom of manufacturing sector can be traced back to the development of various machining operations. The dynamic impact of Machining in the metalworking industry is evident through a wide range of present-day applications. Just in order to help professionals understand the shift of the industrial advancements since the Iron Age era until today's sophisticated technological progressions. These processes are commonly carried out in machine shops or tool room for machining cylindrical or flat jobs to a desired shape, size and finish on a rough block of job material with the help of a wedge shaped tool. The cutting tool is constrained to move relative to the job in such a way that a layer of metal is removed in the form of a chip.

Advantages of Machining process

- A high surface finish can be obtained.
- Machining is not only performed on the metal but it also performs on wood, plastic, composites, and ceramics.
- A variety of geometry features are possible, such a Screw threads, Very straight edges, accurate round holes, etc.
- Good dimensional accuracy.

Limitations of Machining Process

- The accuracy of the components produce is dependent on the efficiency of the operator.
- Consistency in manufacturing is not present. Hence 100% inspection of the component is required.
- The personal needs of the operator are reducing the production rates.
- Because of the large amount of Manpower involved, the labor problem will also be high.

Applications of Machining Process

- The machining can be performed on various components in the form of either conventional or unconventional processes.
- The machining can be performed on a lathe machine, milling machine, ultrasonic machining, etc.
- The advancement of machining can be performed on CNC Machines where there is no intervention of humans.

Manufacturing Cycle

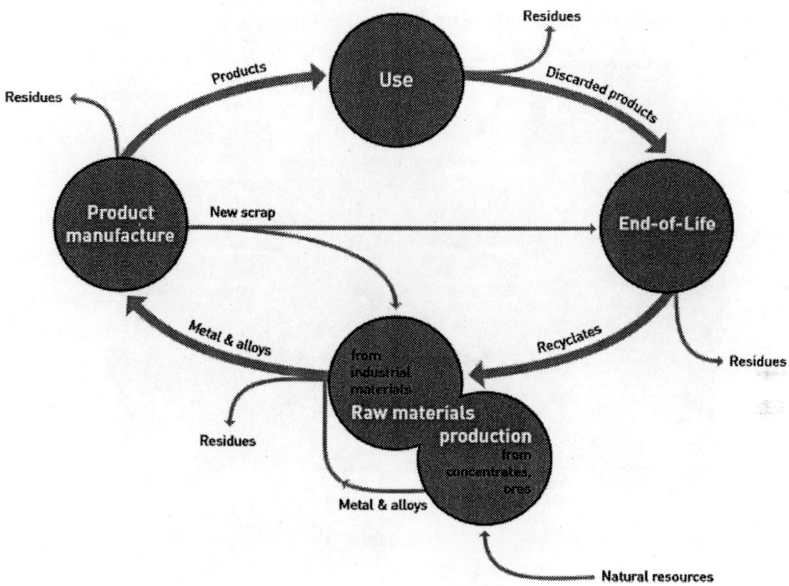

Figure 4.2: Machining process in manufacturing product life cycle.

4.2 Lathe Machine (Turning machine)

Lathe is one of the most versatile and widely used machine tools all over the world. It is commonly known as the mother of all other machine tool. The main function of a lathe is to remove metal from a job to give it the required shape and size. The job is secure1y and rigid1y held in the chuck or in between centers on the lathe machine and then turn it against a single point cutting tool which wi1l remove meta1 from the job in the form of chips.

Figure 4.3: Major parts of a lathe machine

Bed: It is cast into a single piece to serve as a foundation and hence to support remaining parts of thelathe. The top of the bed is made perfectly smooth and flat. Two V-shaped guide ways are machined over it. The outer one guides longitudinally the cutting tool holding unit called **carriage**. The inner one aligns the **live centre** and the **dead centre** in line.

Head stock: It is permanently fixed on the left hand side of lathe bed. It supports the spindle of themachine with the help of bearings on both the sides. It also houses the mechanism to drive the spindle.

Tail stock: It is mounted on the inner guide ways on the right hand side of the lathe bed. It can bemoved on these guide ways towards and away from the head stock and can be clamped at desired position. The spindle inside the tail stock can be moved irrespective of the tail stock. It serves as a support for the workpiece when turning operation is carried out and also houses the tools like drill, reamer, taps, etc.

Carriage: The carriage is mounted on the outer guide ways and moves between head stock and tailstock. It helps in feeding the cutting tool against the workpiece. Also it houses the mechanism for automatic feeding of the cutting tool.

The major parts related with the carriage are:

(a) Saddle: It is the base of the carriage which slides along the ways of the lathe bed in adirection parallel to the lathe axis.

(b) Cross Slide: It is mounted on the saddle. It slides on the guide ways provided on the saddlein a direction perpendicular to the lathe axis.

(c) Compound rest: It is mounted on the cross slide. It has a graduated base and can beswiveled around the vertical axis. The tool can be fed in angular manner.

(d) Tool Post: It is mounted on the compound rest and can slide in a T-slot. A cutting tool canbe firmly held in it.

(e) Lead Screw: It is used in threading operation only. It can rotate by giving spindle rotationsto it. These rotations can be further converted into linear motion by split nut mechanism.

(f) Apron Mechanism: It is an assembly of gears and clutches which converts rotary motion ofthe feed rod into linear motion of the slides. It is used for giving automatic longitudinal or cross feed to the tool.

(g) Feed rod: It also takes rotations from the spindle and converts the rotational motion intolinear motion by apron mechanism. Hence feed rod is used to give automatic longitudinal and cross feed to the slides.

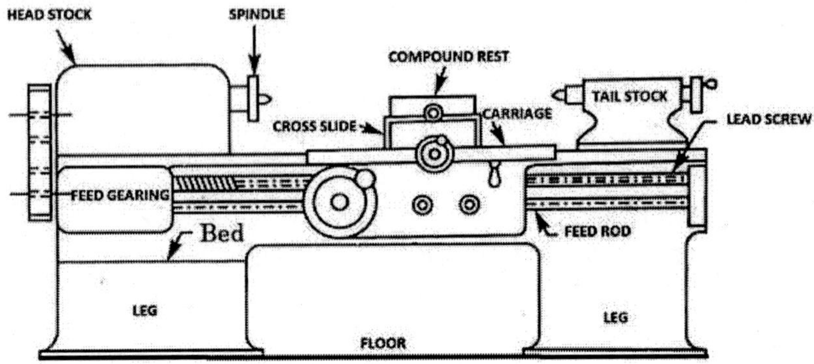

Figure 4.4: Major components of lathe machine.

When the cutting tool moves parallel to workpiece it produces cylindrical job. When the cutting tool moves perpendicular to workpiece it produces flat job. When the cutting tool moves inclined to workpiece it produces tapered surface.

Working Principle of Lathe

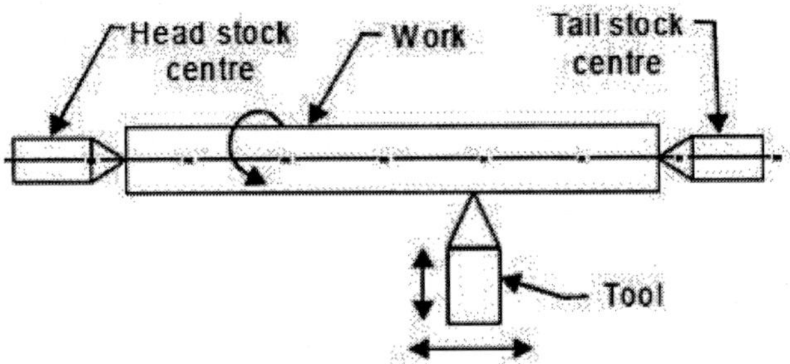

Figure 4.5: Working principle of lathe machine

Specification of Lathe

Lathe specified by one or more of the following criteria

- Distance between centers: it is the maximum length of the job that can be held between the live center and dead center
- Swing diameter: it is the maximum diameter of workpiece that can revolve without touching guideways. Some manufacturers specify

height of centers instead of swing diameter

- Height of centers: it is the height measured from bed to the lathe center axis
- Length of bed: it indicates the approximate floor space occupied by the lathe
- Range of spindle speeds

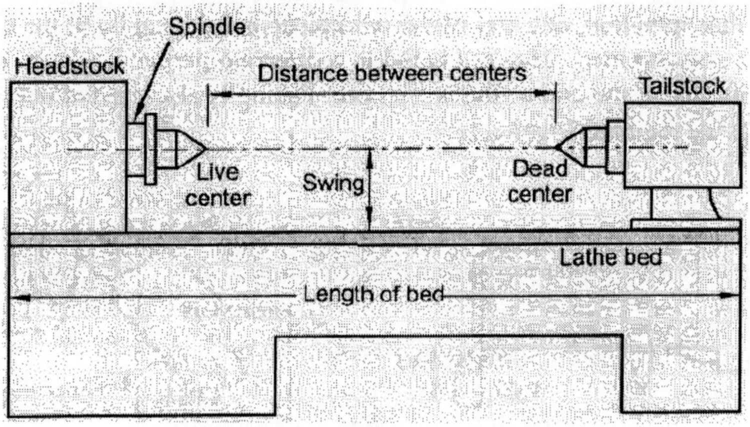

Figure 4.6: lathe specifications

Various Lathe Operations.

The various other operations that you can perform with the help of a Lathe Machine can include sanding, cutting, knurling, drilling, and deformation of tools that are employed in creating objects which have symmetry about the axis of rotation.

Facing

Facing: - Facing is defined as an operation performed on the lathe to generate either flat surfaced or shoulders at the end of the work piece. In facing operation, the direction of feed given is perpendicular to the axis of the lathe. The work piece is held in the chuck and the facing tool is fed either from outer edge of the work piece progressing towards the centre or vice versa. The cutting tool is held by a tool holder in a tool post.

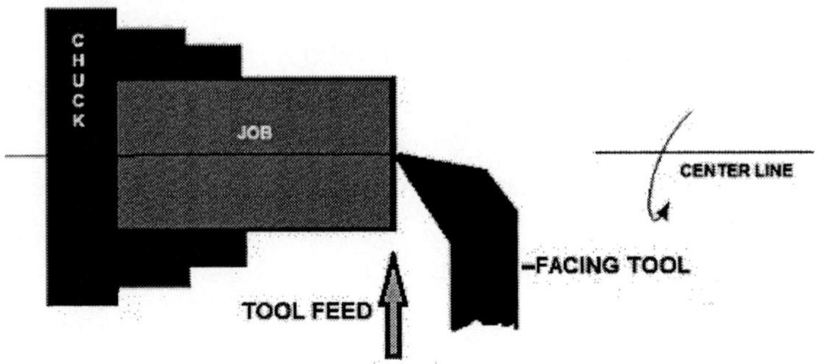

Figure 4.7: Facing Operation

Turning

The work piece is supported in between the two centres which permit the rotation of the work piece. A single point cutting tool is fed perpendicular to the axis of the work piece to a known pre-determined depth of cut, and is then moved parallel to the axis of the work piece. This operation will cut the material which comes out. This method of machining operation in which the work piece is reduced to the cylindrical section of required diameter is called 'Turning' . It is a process of removal of excess material from the workpiece in minimum time by applying a high rate of feed and heavy depth of cut. The depth of cut is around 2 to 5mm and the rate of feed is 0.3 to 1.5mm/revolution.

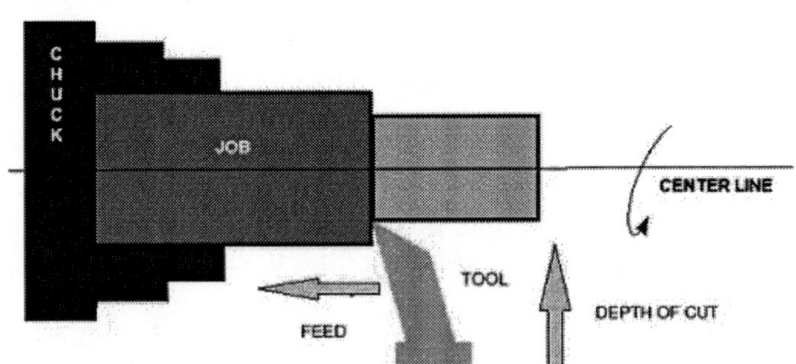

Figure 4.8: Turning Operation

Taper Tuning

A taper is defined as a uniform decrease or increase in the diameter of a workpiece along with its length.

The operation by which a conical surface of the gradual reduction in diameter from a cylindrical workpiece is produced is called taper turning.

The half taper angle is determined by using the formula $\tan a = (D1 - D2)/\,2L$

A tapering form may be done by any one of the following methods.

1. By swiveling the compound rest
2. Tail-stock set over method
3. By taper turning attachment
4. Taper turning by form tool

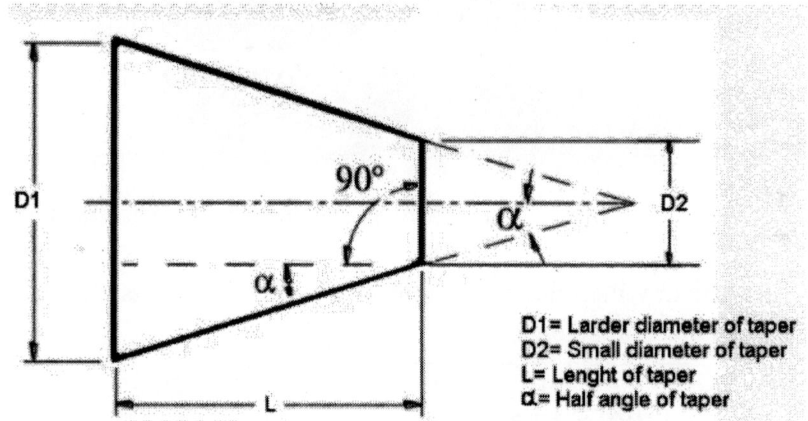

Figure 4.9: Taper Turing operation.

Taper turning by swiveling the compound rest:

The compound tool rest is swivelled to the required taper angle and then locked in the angular position. The carriage is also locked at that position. For taper turning, the compound tool rest is moved linearly at an angle so that the cutting tool produces the tapered surface on the work piece. This method is limited to short tapered lengths due to the limited movement of the compound toolrest.

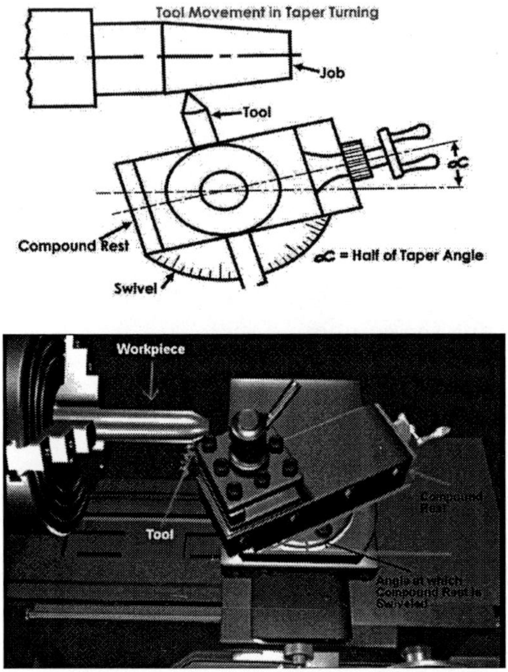

Figure 4.10: Taper Turning by Swiveling the compound rest

The angle at which the compound rest to be swiveled is calculated using the equation given below

$$\tan\alpha = \frac{D-d}{2\times L} \qquad \text{or} \qquad \alpha = \tan^{-1}\frac{D-d}{2\times L}$$

Where

D = larger diameter of taper in mm

d = smaller diameter of taper in mm

L = length of taper in mm

α = half of taper angle in degree

Tail-stock set over (offset) method:

Set over of tail-stock from its center-line is done equal to half taper. Job is held between the centers. The length of the workpiece will be long enough. An only a small taper on a long job is done by this process. It is used for external taper only.

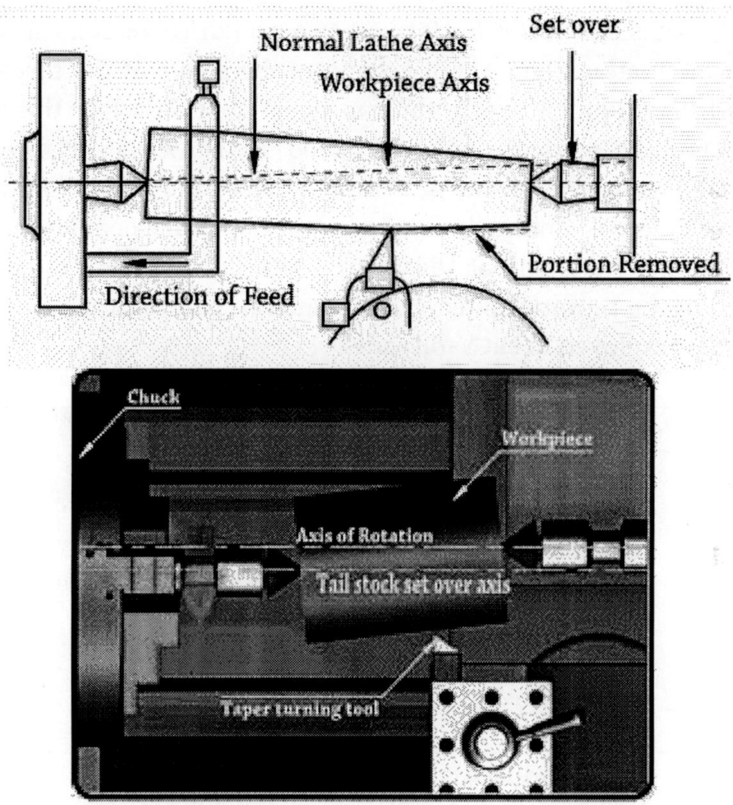

Figure 4.11: Taper Turning by offsetting the compound rest

A thread is a helical ridge formed on a cylindrical or conical rod. It is cut on a lathe when a tool ground to the shape of the thread, is moved longitudinally with uniform linear motion while the work piece is rotating with uniform speed as shown in Fig. By maintaining an appropriate gear ratio between the spindle on which the work piece is mounted, and the lead screw which enables the tool to move longitudinally at the appropriate linear speed, the screw thread of the required pitch can be cut. The pointed tool shown in Fig, is employed to cut V-threads. When square threads are to be cut, the tool is ground to a squared end.

Threading:

A thread is a helical ridge formed on a cylindrical or conical rod. It is cut on a lathe when a tool ground to the shape of the thread, is moved longitudinally with uniform linear motion while the work piece is rotating with uniform speed as shown in Figure. By maintaining an appropriate gear ratio between the spindle on which the work piece is mounted, and the lead screw which enables the tool to move longitudinally at the appropriate linear speed, the screw thread of the required pitch can be cut. The pointed tool shown in Figure is employed to cut V-threads.

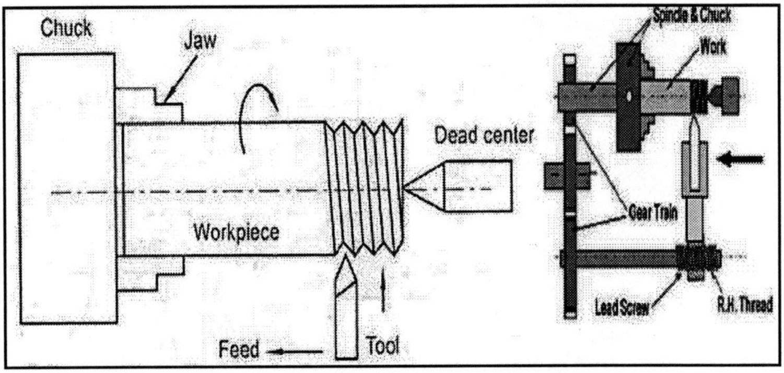

Figure 4.12: Threading on a lathe

Knurling:

knurling is a finishing process used to create any combination of horizontal, vertical or crossing lines on the surface of a workpieceIt is an operation of obtaining a diamond shape on the workpiece for the gripping purpose. This is done to provide a better gripping surface when operated by hands. It is done using a knurling tool. The tool consists of a set of hardened steel roller, and it is held rigidly on the toolpost.

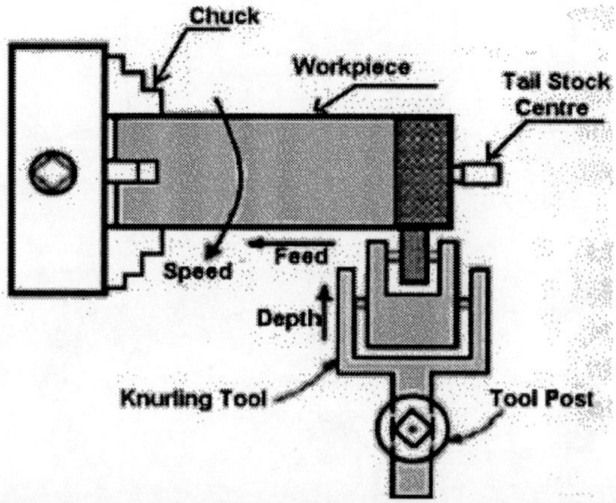

Figure 4.13: Knurling operation

Drilling

Drilling is the operation of producing circular holes on object by means of twist drill bit, In this operation drill tool is holds in the tool post and glides on the bed, whereas work piece is placed in 3 jaw chuck and it is continuously rotating the drilled hole depth can be controlled by tailstock hand wheel for moving sleeve. Drilling can perform only at one end of the work piece.

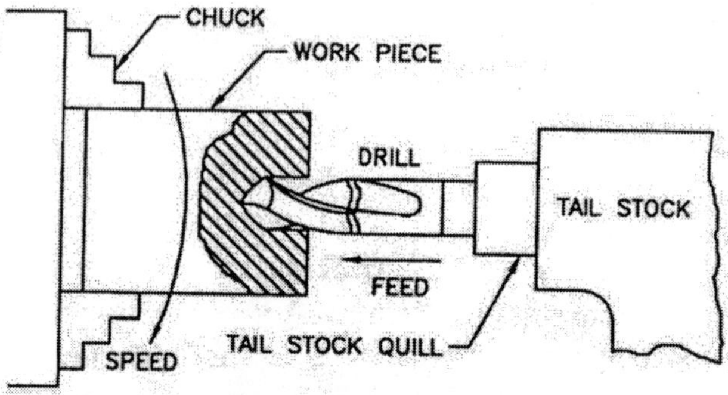

Figure 4.14: Drilling operation

4.3 Milling:

Milling is a manufacturing process in which the excess material from the work piece is removed by a rotating multipoint cutting tool called milling cutter. The milling cutter is a multipoint cutting tool. The work piece is mounted on a movable work table which will be fed against the revolving milling cutter to perform the cutting operation.

Types of Milling Machines

Milling machines are a type of machinery for removing material from a workpiece using rotary cutters. These machines can drill, bore, and cut an array of materials. This process of removing pieces of the material in line with the tool axis is known as milling, and thus the machines used in the process are called milling machines. Milling machines come in many types and are used across a variety of industries.

Following are the types of Milling Machines used at commercial grade

- Column Milling Machines

Often used for creating car parts, a column milling machine is one of the simplest types of milling machine. They consist of 5 key parts the worktable, head, saddle, knee, and over arm and use a vertically suspended drill.

- Horizontal Milling Machines

Named as it is positioned horizontal to the ground, horizontal milling machines work by moving the bench the workpiece is placed on sideways whilst the cutting tool moves vertically.

Parts of Milling Machines

Figure 4.15: Horizontal and Vertical Milling machines

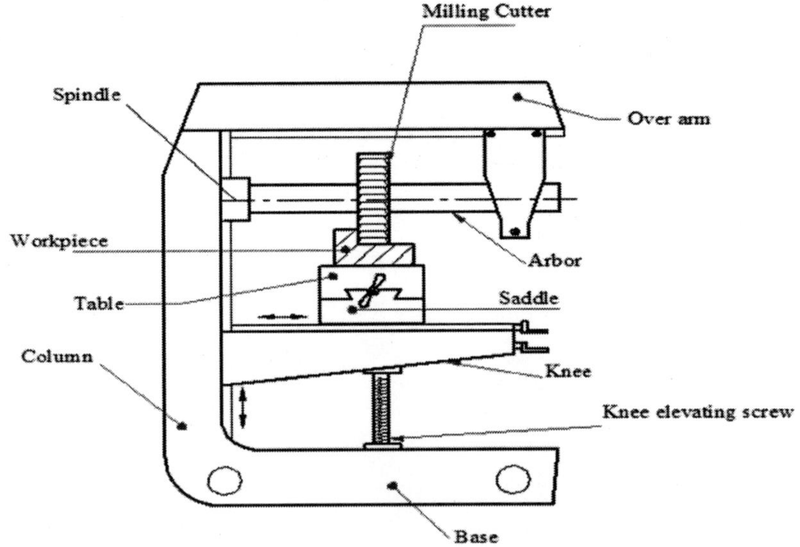

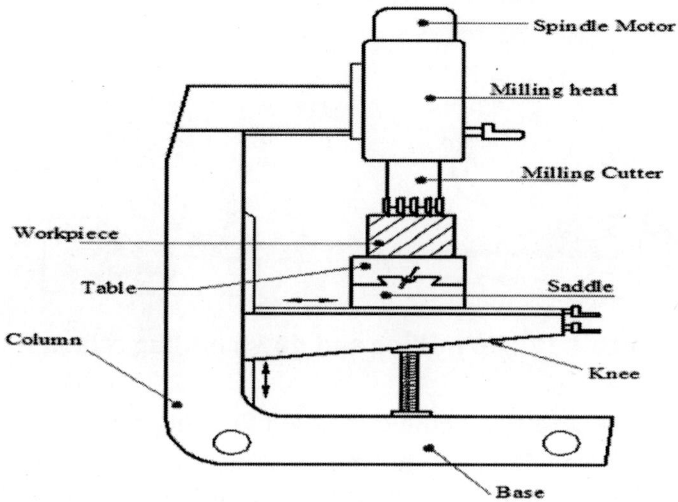

Figure 4.16: Schematics of Horizontal and Vertical Milling machines

Parts of Milling Machine

1. **Base:** It is a heavy casting and supports all other parts.

2. **Column:** It is box shaped main supporting frame located at one end of the base and is having dovetailguide ways on the front face. It houses all the driving mechanism for spindle.

3. **Knee:** It is capable of sliding up and down on the guide ways of the column. The adjustment is done by

operating the elevating screw. It houses the feed mechanism for table. The top face is the slide way for the saddle to provide cross travel of the table.

4. **Saddle:** This is placed on the top of knee. The top face provides guide ways for the table.

5. **Table:** This rests on the ways of the saddle. It is having T-slots. Longitudinal travel is provided through lead screw. It moves perpendicular to the movements of saddle.

6. **Spindle:** It is vertical or perpendicular to the table. The cutter is fixed to one end of the spindle and rotates about a vertical axis in the horizontal plane

Principle of Milling: Up Milling and Down Milling:

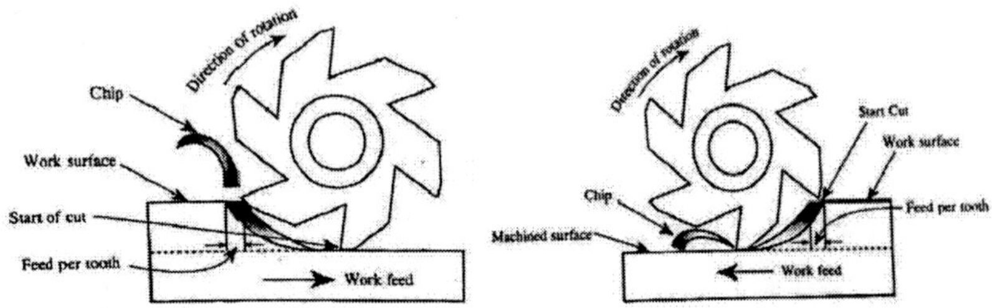

Figure 4.17: Up milling and down milling schematic

The above figure shows the principle of milling process that is up milling and down milling. The up milling is also called as conventional milling and down milling is called climb milling. The milling cutter is mounted on a rotating shaft known as arbor. The work piece which is mounted on the table can be fed either in the direction opposite to that of the rotating cutter as shown in figure a or in the same direction of the cutter as shown in figure.

Up Milling	Down Milling
The cutter rotates against the direction of travel of the work piece.	The cutter rotates in the same direction of travel of the work piece.
The cutter rotates in the same direction of travel of the work piece.	The chip thickness is maximum at the beginning of the cut and or reduces to a minimum when the cut terminates.
The cutting force is zero at the beginning of the cut and maximum at the end of the cut.	The cutting force is maximum when the cut begins and it reduces to a minimum when the tooth leaves the work.
The cutting force is	The cutting force is

directed upward and this tends to lift the work piece from the fixture.	downwards which tends to seat the work piece firmly in the fixture.
Difficulty in pouring the coolant just on the cutting edge, due to the typical nature of the cut.	The coolant can be poured directly at the cutting zone where cutting force is maximum.
The coolant can be poured directly at the cutting zone where cutting force is maximum.	Chips are disposed of easily and do not interfere with the cutting.
The actual cutting does not start at zero chip thickness as teeth slide through a minute distance which gives a slightly wavy surface.	Chips are disposed of easily and do not interfere with the cutting.

Milling Operations

i) plane milling

Plane Milling or Slab Milling

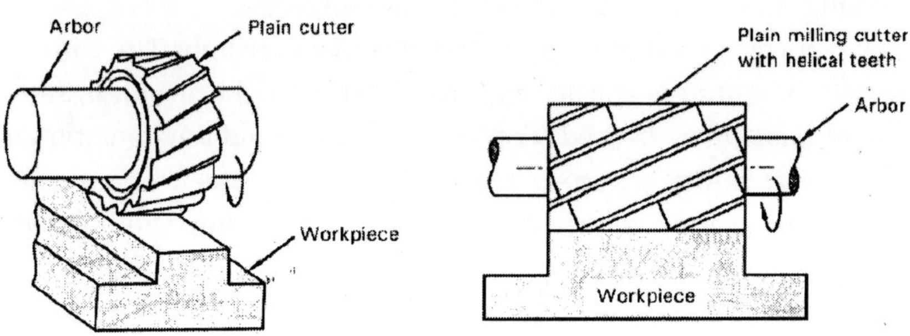

Figure 4.18: Plain milling schematic

Plane milling also called surface milling or slab milling in the operation producing a plane horizontal surface with a miller cutter whose axis is parallel to the surface of workpiece being machined. The cutter called plain or slab milling cutter is mounted on arbor of a horizontal milling machine to carry out the machining operation, the cutter has straight or helical teeth on the peripheral cylindrical surface, where as helical tooth cutter removes greater amount of material then straight tooth cutter

End Milling

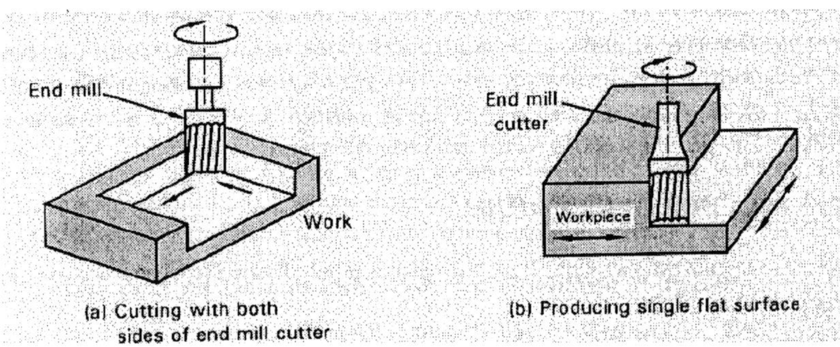

Figure 4.19: End milling schematic

End milling is operation performed for producing flat surfaces, slots, grooves or finishing the edges of the workpiece by means of a tool called end mill or end milling cutter. The tool has teeth at ends and as well as periphery (Sides) and hence can be configured to cut with both its ends and the side as shown in figure

A flat surface produced by the end side of cutter only. The cutter is typically mounted vertically and may have straight or spiral flutes. Straight flute cutter is used for milling both soft and tough materials, while spiral flutes are used mostly for cutting steel. The surface produced by end mill may horizontal, vertical or inclined with respect to top of machine table

T-slot Milling

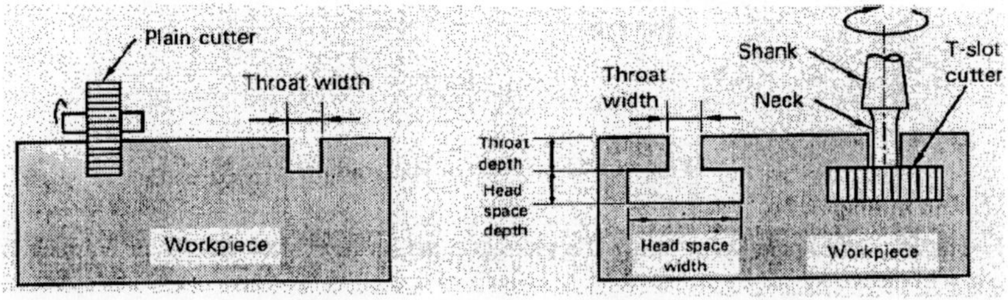

Figure 4.20: Slot milling schematic

Slot milling is operation of producing various type of slots like T-slots, plane slots, dovetail slots etc., in workpieces the type of cutter selected depends on the shape of the slot to be produced.

Producing T-slots

Two separate milling cutters are required on milling T-slots. Initially a side cutter or an ed milling cutter is used to cut throat (open slot) starting from one end of the workpiece to its other end. A T-slot milling cutter is then used to cut the head space to the desired dimensions.

Form Milling

Form milling is the operation of producing curved profiles with a verity of shapes like concave convex spline etc., using cutters whose edge is shaped to produce a special configuration on the surface of the workpiece. From figure the cutter known as form mill has teeth on its periphery and designed in various shapes to suit the type of surface being machined. Form mill is the gear tooth cutter, which is used to cut gear teeth on workpieces

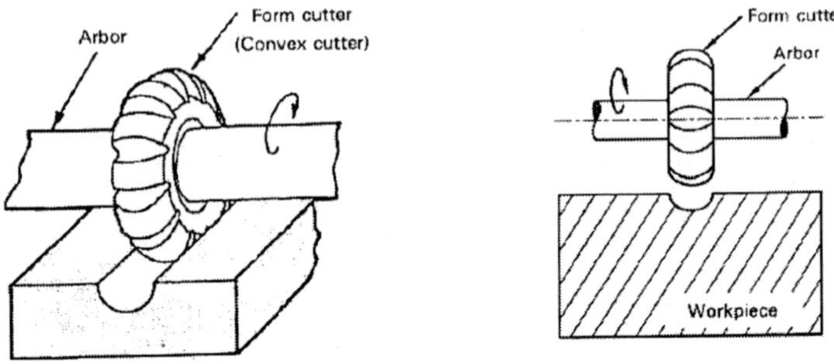

Gang Milling

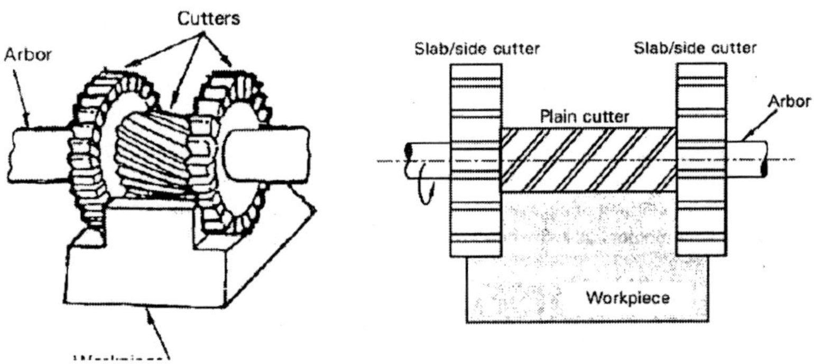

Figure 4.21: Up milling and down milling schematic

Gang milling is the operation in which two or more cutters are mounted on the same arbor, so that different profile required on the work piece can be machined simultaneously in a single pass. Following figure shows the gang milling operation. All the cutter used may be of same type or of different types and it depends on the type of surface being machined.

4.4 Computer Numerical Control (CNC)

Computer Numerical Control (CNC) is a method of automating control of machine tools through the use of software embedded in a microcomputer attached to the tool.With CNC, each object to be manufactured gets a custom computer program, usually written in an international standard language called G-code, stored in and executed by the machine control unit(MCU).The program contains the instructions and parameters the machine tool will follow,such as the feed rate of materials and the positioning and speed of the tool's components.Mills,lathes,routers,laser cutters are common machine tools whose operations can be automated with CNC. It can also be used to control non-machine tools,such as welding, electronic assembly machines. CNC is considered to provide more precision, complexity and repeatability than is possible with manual machining. Other benefits include greater accuracy, speed and flexibility, as well as capabilities such as contour machining, which allows milling of contoured shapes, including those produced in 3D designs.

2.16 Components of CNC

The main components of CNC are:

1. **Input Devices**: These are the devices which are used to input the part program in the CNC machine. There are three commonly used input devices and these are punch tape reader, magnetic tape reader and computer via RS-232-C communication.
2. **Machine Control Unit (MCU)**: It is the heart of the CNC machine. It performs all the controlling action of the CNC machine.
3. **Machine Tool**: A CNC machine tool always has a slide table and a spindle to control of the position and speed. The machine table is controlled in X and Y axis direction and the spindle is controlled in the Z axis direction.Depending upon the operation to be performed,the tool can be selected from tool turret.Major type of tools used in milling operation are shown in the second picture.

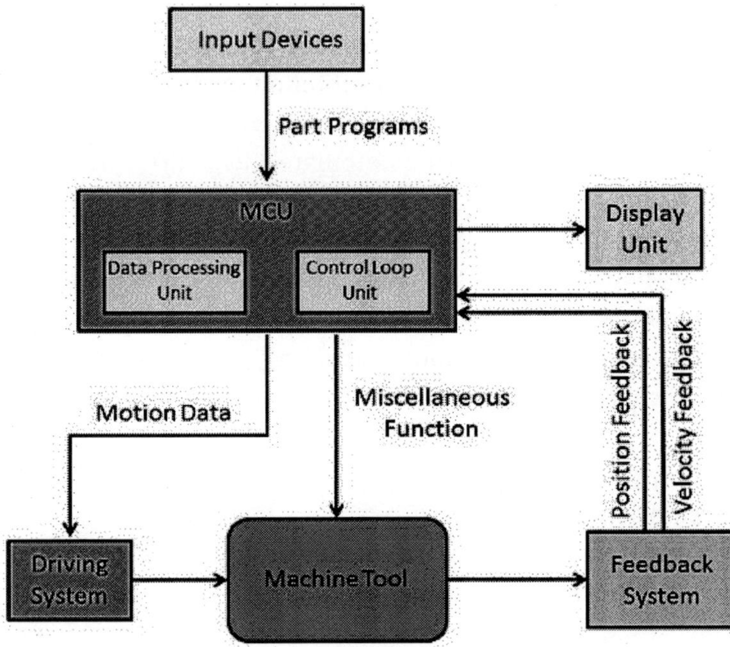

Figure 2.22: Components of CNC system.

4.5 Working of CNC Machine

First, the part program is inserted into the MCU of the CNC. In MCU all the data process takes place and according to the program prepared, it prepares all the motion commands and sends it to the driving system.The drive system works as the motion commands are sent by MCU. The drive system controls the motion and velocity of the machine tool.The feedback system records the position and velocity measurement of the machine tool and sends a feedback signal to the MCU. In MCU, the feedback signals are compared with the reference signals and if there are errors, it corrects it and sends new signals to the machine tool for the right operation to happen.A display unit is used to see all the commands, programs and other important data. It acts as the eye of the machine.

Advantages of CNC machines

1. CNC Machining Produces Little to No Waste
2. Zero Defects and Greater Accuracy
3. Faster and Efficient Production
4. Quicker Assembly
5. Enhanced Personnel Safety
6. Reduction in Energy Consumption
7. CNC Machining Leads to Lower Production Costs

Applications of CNC machines

1. Machining is accurate
2. Time taken to perform a job is very less
3. Safe to operate
4. Number of operators required to operate a machine are reduced
5. No possibility of human error
6. Reliable
7. Even very complex designs can also be made
8. Low maintenance required
9. They are versatile
10. Uniformity in designs
11. They could run for all 24 hours a day

CNC Machining centres

The CNC machine tool center is an advanced manufacturing machine tool that can perform a variety of machining operations with high precision, high quality and high surface finish. A CNC machine tool center can perform drilling, milling and lathe operationsThe manufacture of prismatic parts in the industry, such as gearboxes, partitions, frames, covers, etc., requires different types of operations such as milling, boring, drilling, tapping and many other related machining operations. In the past, this production process had to be divided into many working stages, and the operation on different machine tools was able to produce a finished product, which was resulted in a large amount of delivery time and cost. In order to overcome this problem, a CNC machine tool center was developed. Milling, lathing and drilling operations on a single machine tool allows one machine to perform a greater variety of machining needs.

CNC Turning centres

CNC turning centers and lathes may seem identical, but they generally refer to slightly different machines. Lathes generally only turn while turning centers can be more advanced. These machines can be capable of facing, threading, knurling, drilling, boring, reaming, and taper turning. Horizontal CNC turning machines are more common, but vertical machines utilize gravity. Between CNC turning, CNC turning centers, the different types of CNC machines.

CNC machining is far better than manual machining for a few specific reasons:

- Accuracy: Programming a CNC to complete a job creates a consistent and accurate cut to ensure product quality. It can be repeated over and over in the exact same manner.
- Cost Effectiveness: The cost of hiring a person to manually machine parts at scale is far more expensive than having a machine do it, as a machine will not tire, does not need lunch breaks, and can, theoretically run 24/7.
- Speed: CNC turning can churn out hundreds and hundreds of parts while retaining a consistent end product.

4.6 Part Program for CNC Turning Operations

The part program is a sequence of instructions, which describe the work, which has to be done on a part, in the form required by a computer under the control of computer numerical control (CNC) software. It is the task of preparing a program sheet from a drawing sheet. All data is fed into the CNC system using a standardized format. Programming is where all the machining data are compiled and where the data are translated into a language which can be understood by the control system of the machine tool.

The machining data is as follows:

1. Machining sequence classification of process, tool start up point, cutting depth, tool path, etc.
2. Cutting conditions, spindle speed, feed rate, coolant, etc.
3. Selection of cutting tools.

Structure of part program

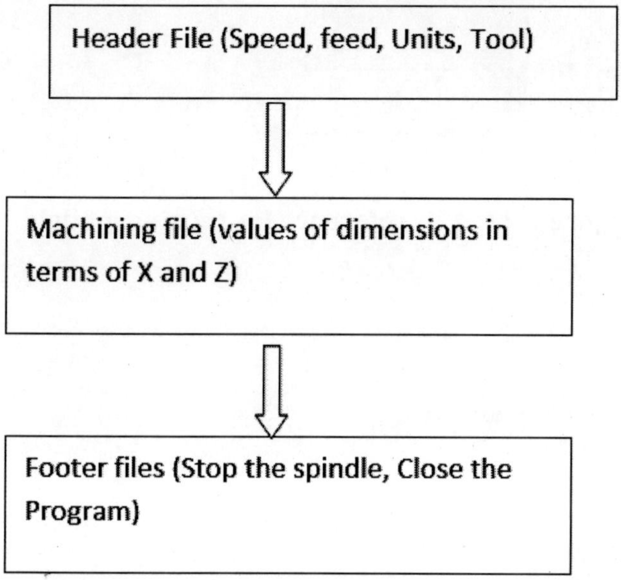

Figure 2.23: General Structure of a CNC Program.

G codes and M Codes used in the program

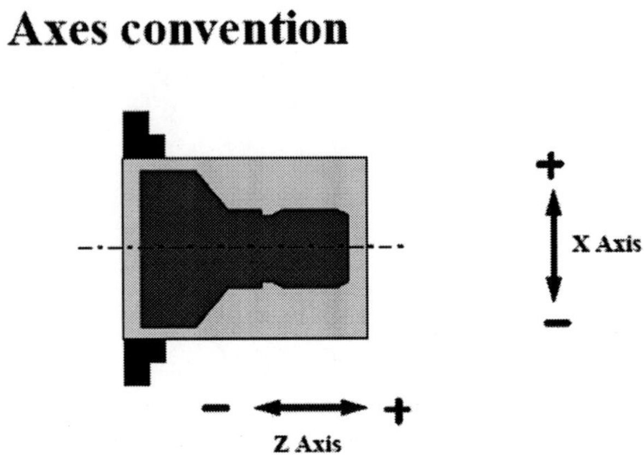

Figure 2.24: Axis reference for turning models.

List of G-codes

G-code	Function
G00	Positioning rapid traverse
G01	Linear interpolation (feed)
G02	Circular interpolation CW
G03	Circular interpolation CCW
G04	Dwell
G20	Inch unit
G21	Metric unit
G28	Automatic zero return
G30	2nd reference point return
G32 / G33	Thread cutting (single motion)
G40	Tool nose radius compensation cancel
G41	Tool nose radius compensation left
G42	Tool nose radius compensation right
G50 / G92	Local co-ordinate system setting, max. spindle speed setting
G70	Finishing cycle
G71	Stock removal in turning
G72	Stock removal in facing
G73	Pattern repeating
G74	Peck drilling on Z axis / Face grooving
G75	Peck drilling on X axis / Int-Ext grooving
G76	Multiple threading cycle
G94 / G98	Feed per minute
G95 / G99	Feed per revolution
G96	Constant surface speed control
G97	Constant surface speed control cancel

List of M codes

M codes vary from machine to machine depending on the functions available on it and the manufacturer of the machine decides them. The M codes listed below are the common ones.

M-codes	Function
M00	Optional program stop automatic
M01	Optional program stop request
M02	Program end
M03	Spindle ON clock wise (CW)
M04	Spindle ON counter clock wise (CCW)
M05	Spindle stop
M06	Tool change
M07	Mist coolant ON (coolant 1 ON)
M08	Flood coolant ON (coolant 2 ON)
M09	Coolant OFF
M30	End of program, Reset to start
M98	Sub program call
M99	Sub program end

There are other M-codes for functions like gear change, tail stock quill in/out, chuck clamp/unclamp, chip conveyor forward/backward, door open / close etc.

Program 1: Prepare a part program for the geometry given below.

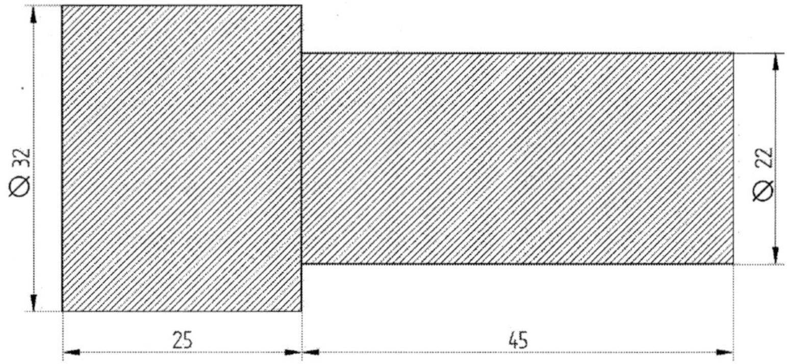


```
O002
N001    G21    G98
N020    G28    U0     W0
N030    M03    S1200
N040    M06    T0101
N050    G00    X35    Z2
N060    G90    X32    Z-45    F50
```

```
N070              X31
N080              X30
N090              X29
N100              X28
N110              X27
N120              X26
N130              X25
N140              X24
N150              X23
N160              X22
N170      G28     U0      W0
N180      M05
N190      M30
```

Program 02: Prepare a part program for the geometry given below.

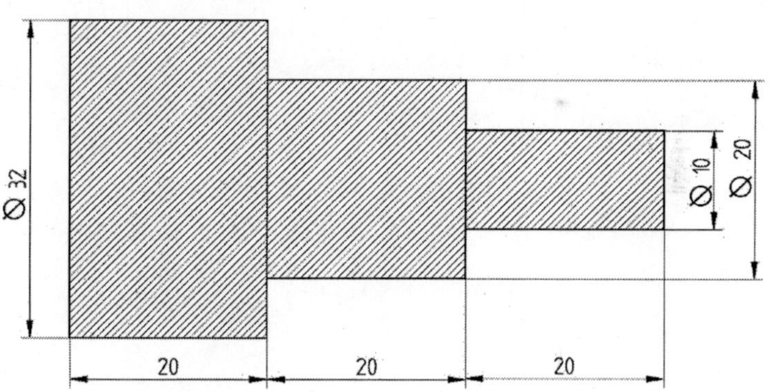


```
O001
N001      G21     G98
N002      G28     U0      W0
N003      M03     S1000
N004      M06     T0101
N005      G00     X32     Z2
N006      G90     X30     Z-40    F50
N007              X29
```

```
N008            X28
N009            X27
N010            X26
N011            X25
N012            X24
N013            X23
N014            X22
N015            X21
N016            X20
N017            X19    Z-20
N018            X18
N019            X17
N020            X16
N021            X15
N022            X14
N023            X13
N024            X12
N025            X11
N026            X10
N027      G28   U0     W0
N180      M05
N190      M30
```

Program 03: Prepare a part program for the geometry given below

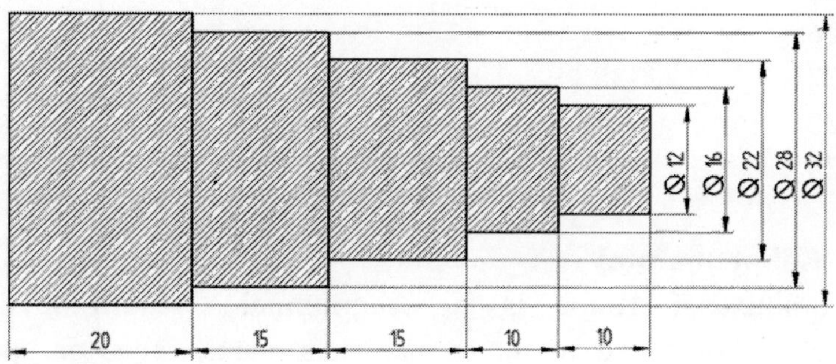


```
O003
N001    G21    G98
N002    G28    U0     W0
N003    M03    S5000
N004    M06    T0101
N005    G00    X33    Z2
N006    G90    X32    Z-50    F40
N007           X31
N008           X30
N009           X29
N010           X28
N011           X27    Z-35
N012           X26
N013           X25
N014           X24
N015           X23
N016           X22
N017           X21    Z-20
N018           X20
N019           X19
N020           X18
N021           X17
N022           X16
```

N023		X15	Z-10
N024		X14	
N025		X13	
N026		X12	
N027	G28	U0	W0
N028	M05		
N029	M30		

4.7 Robot anatomy

An industrial robot is a programmable, multi-functional manipulator designed to move materials, parts, tools, or special devices through variable programmed motions for the performance of a variety of tasks.

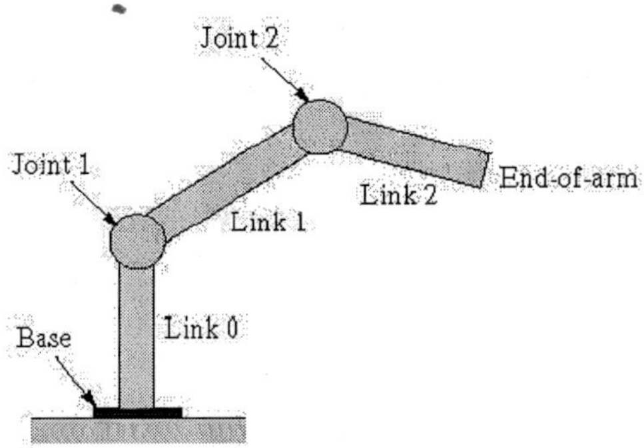

Figure 4.25: Robot Anatomy

An industrial robot consists of a mechanical manipulator and a controller to move it and

Perform other related functions

- The mechanical manipulator consists of joints and links to position and orient the endof the manipulator relative to its base
- The controller operates the joints in a coordinated fashion to execute a programmedwork cycle
- A robot joint is similar to a human body joint it provides relative

movement between two parts of the body

• Typical industrial robots have five or six joints, Manipulator joints: classified as linearor rotating

Robotic Joints

1. Based on Translational motion

➢ Linear joint (type L)
➢ Orthogonal joint (type O)

2. Based on Rotary motion

➢ Rotational joint (type R)
➢ Twisting joint (type T)
➢ Revolving joint (type V)

a) Linear joint (type L joint)

The relative movement between the input link and the output link is a translational sliding motion, with the axes of the two links being parallel.

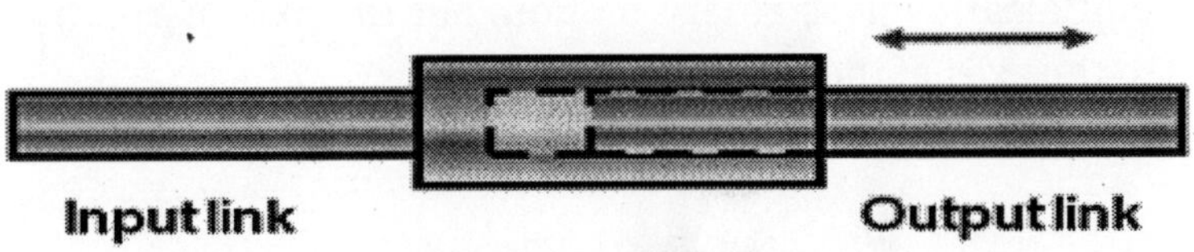

(a) Linear Joint

b) Orthogonal joint (type U joint)

This is also a translational sliding motion, but the input and output links are perpendicular to each other during the move

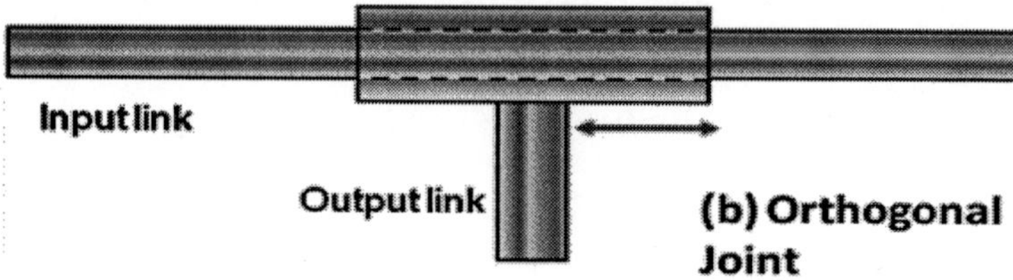

(b) Orthogonal Joint

c) Rotational joint (type R joint)

This type provides rotational relative motion, with the axis of rotation perpendicular to the axes of the input and output links

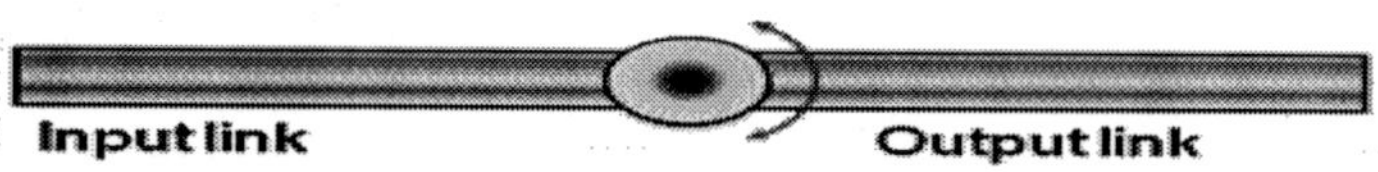

(c) Rotational Joint

d) Twisting joint (type T joint)

This joint also involves rotary motion, but the axis or rotation is parallel to the axes of the two links.

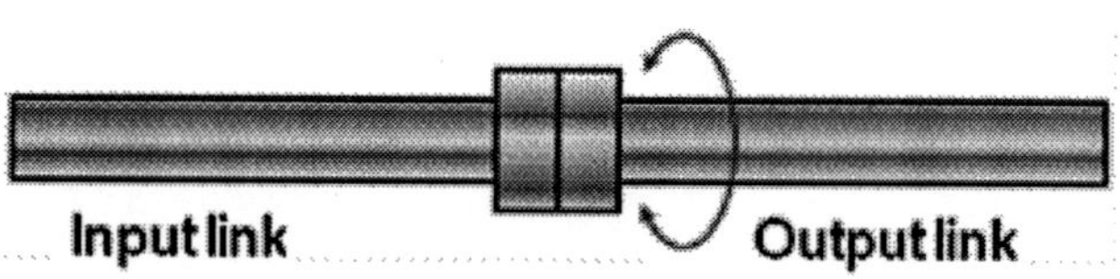

(d) Twisting Joint

e) Revolving joint (type V-joint, V from the "v" in revolving)

In this type, axis of input link is parallel to the axis of rotation of the joint. However the axis of the output link is perpendicular to the axis of rotation.

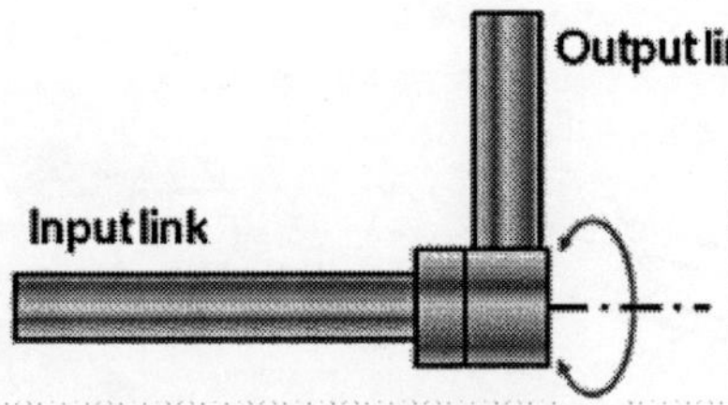

(e) Revolving Joint

Classification of robots Based on Physical Configuration

The following are the important physical configuration of industrial robots

1. **Polar configuration**
2. **Cylindrical configuration**
3. **Cartesian co-ordinate robot**
4. **Jointed-arm robot**
5. **SCARA**

Polar configuration:

- This configuration also goes by the name "spherical coordinate" because the workspace within which it can move its arm is a partial sphere as shown in figure. The robot has a rotary base and a pivot that can be used to raise and lower a telescoping arm

- It consists of a sliding arm L-joint, actuated relative to the body, which rotates around both a vertical axis (T-joint), and horizontal axis (R-joint).

- The work volume of polar configuration is hemisphere

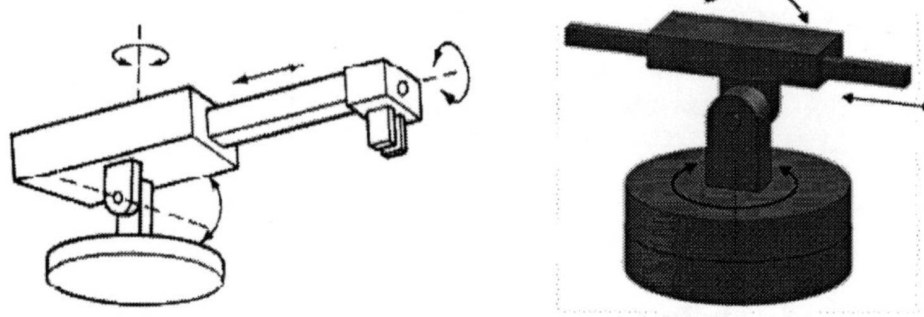

Figure 4.26: Polar Configuration

Cylindrical configuration:

- ➢ In this configuration, the robot body is a vertical column that swivels about a vertical axis.
- ➢ The arm consists of several orthogonal slides which allow the arm to be moved up or down and in and out with respect to the body. This is illustrated schematically in figure.
- ➢ Common configuration is to use a T-joint to rotate the column about its axis. An L-joint is used to move the arm assembly vertically along the column, while an O-joint is used to achieve radial movement of the arm.
- ➢ Work volume of this robot is cylindrical

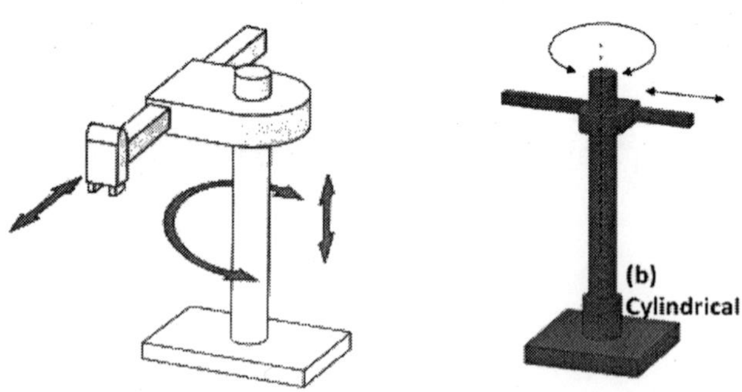

Figure 4.27: Cylindrical Configuration

Cartesian configuration:

➢ A robot which is constructed around this configuration consists of three orthogonal slides, as shown in fig. the three slides are parallel to the x, y, and z axes of the Cartesian coordinate system.

➢ It is also known as rectilinear robot and x-y-z robot. It consists of three sliding joints, two of which are orthogonal O-joints.

➢ By appropriate movements of these slides, the robot is capable of moving its arm at any point within its three dimensional rectangular spaced work space.

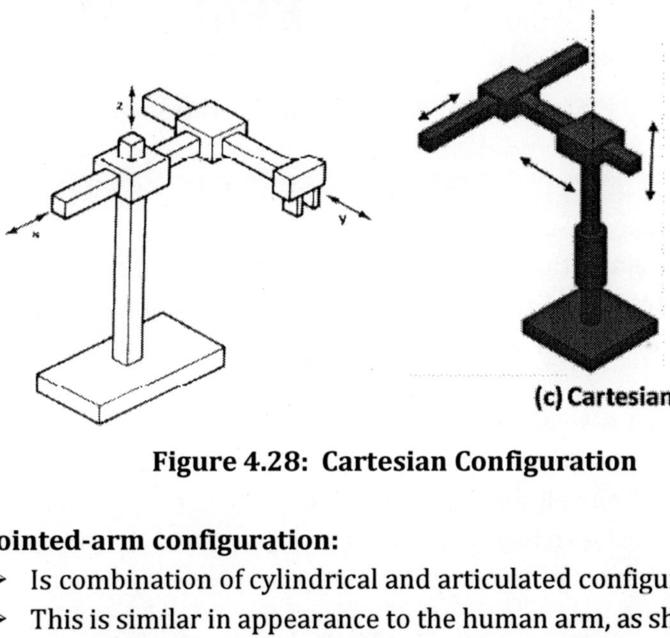

Figure 4.28: Cartesian Configuration

Jointed-arm configuration:

➢ Is combination of cylindrical and articulated configurations?

➢ This is similar in appearance to the human arm, as shown in fig. the arm consists of several straight members connected by joints which are analogous to the human shoulder, elbow, and wrist.

➢ The robot arm is mounted to a base which can be rotated to provide the robot with the capacity to work within a quasi-spherical space

➢ It is similar to the configuration of a human arm. It consists of a vertical column that swivels about the base using a T-joint. Shoulder joint (R-joint) is located at the top of the column. The output link is an elbow joint (another R joint).

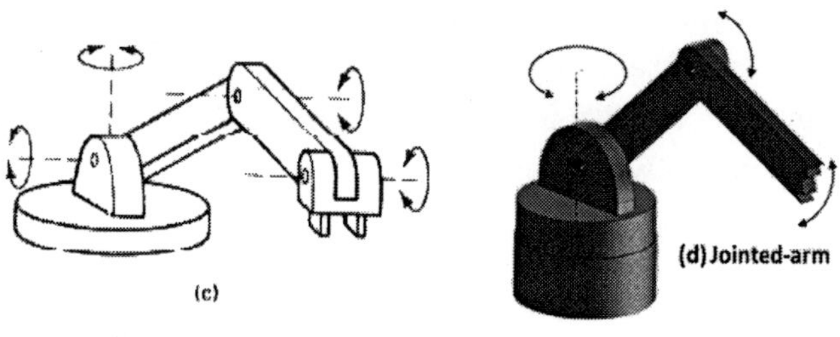

Figure 4.29: Jointed Configuration

SCARA

Its full form is 'Selective Compliance Assembly Robot Arm'. It is similar in construction to the jointer-arm robot, except the shoulder and elbow rotational axes are vertical. It means that the arm is very rigid in the vertical direction, but compliant in the horizontal direction.

Robot wrist assemblies consist of either two or three degrees-of-freedom. A typical three-degree-of-freedom wrist joint is depicted in Figure 7.5.4. The roll joint is accomplished by use of a T-joint. The pitch joint is achieved by recourse to an R-joint. And the yaw joint, a right-and-left motion, is gained by deploying a second R-joint.

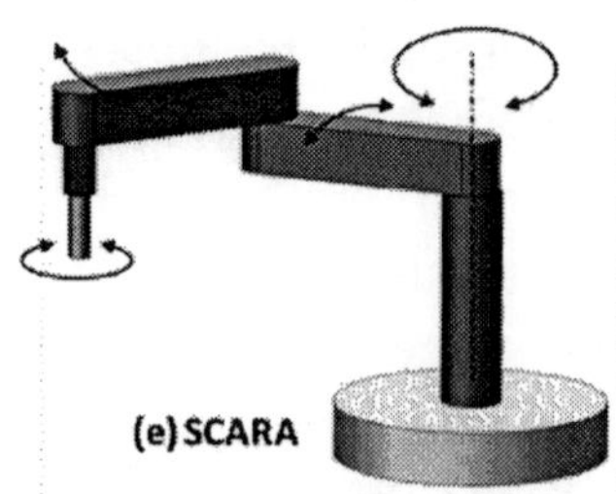

Figure 4.30: SCARA Configuration

End Effectors

- In the terminology of robotics, end effectors can be defined as a device which is attached to the robots wrist to perform a specific task.
- The task might be work part handling, spot welding, spray painting, or any of a great variety of other functions.
- The possibilities are limited only by the imagination and ingenuity of the application engineers who design robot systems.
- The end effectors are the special purpose tooling which enables the robot to perform a particular job. It is usually custom engineered for that job, either by the company that owns the robot or company that sold the robots.
- Most robot manufacturer has engineered groups which design and fabricate end effectors or provide advice to their customers on end effectors design.
- For purpose organization, we will divide the various types of end effectors into two categories: grippers and tools.

1. Grippers: are generally used to grasp and hold an object and place it at a desired location. Grippers can be classified as

- Mechanical grippers
- Vacuum or suction cups
- Magnetic grippers
- Adhesive grippers
- Hooks,
- Scoops, and so forth.

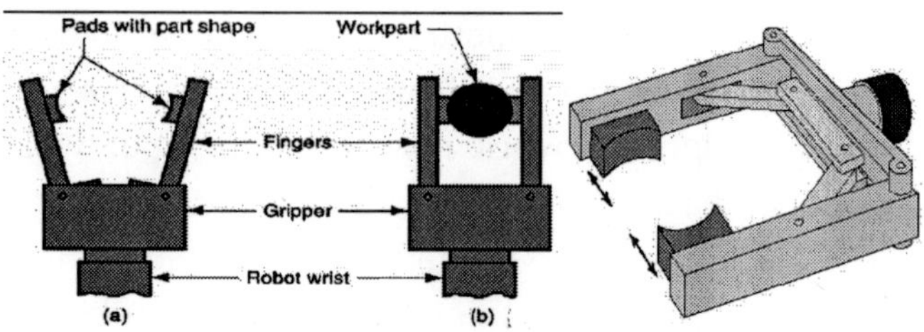

S

Figure 4.31: End Effector

2. Tools: a robot is required to manipulate a tool to perform an operation on a work part. Here the tool acts as end-effectors. Spot-welding tools, arc-welding tools, spray painting nozzles, and rotating spindles for drilling and grinding are typical examples of tools used as end-effectors.

➢ The camera is mounted either on the robot or in a fixed position above the robot so that its field of vision includes the robots work volume. The computer software enables the vision system to sense the presence of an object and its position and orientation.

➢ Vision capability would enable the robot to carry out the following kinds of operations. Retrieve parts which are randomly oriented on a conveyor Recognize particular parts which are intermixed with other objects Perform assembly operations which require alignment

Sensors used in Robots

For certain robot application, the type of workstation control using interlocks is not adequate the robot must take on more human like senses and capabilities in order to perform the task in a satisfactory way these senses and capability includes vision and hand eye coordination, touch, hearing accordingly we will divide the types of sensors used in robotics into the following three categories.

1. Vision sensors
2. Tactile and proximity sensors
3. Voice sensors

Vision sensors

➢ This is one of the areas that is receiving a lot of attention in robotics research computerized visions systems will be an important technology in future automated factories.

➢ Robot vision is made possible by means of video camera a sufficient light source and a computer programmed to process image data.

➢ The camera is mounted either on the robot or in a fixed position above the robot so that its field of vision includes the robots work volume. The computer software enables the vision system to sense the presence of an object and its position and orientation.

➢ Vision capability would enable the robot to carry out the following kinds of operations. Retrieve parts which are randomly oriented on a conveyor Recognize particular parts which are intermixed with other objects Perform assembly operations which require alignment

Tactile and proximity sensor

Tactile sensors provide the robot with the capability to respond to contact forces between

Itself and other objects within its work volume. Tactile sensors can be divided into two types:

1. Touch sensors
2. Stress sensors

➢ **Touch sensors** are used simply to indicate whether contact has been made with an object. A simple micro switch can serve the purpose of a touch sensor. Stress sensors are used to measure the magnitude of the contact force. Strain gauge devices are typically employed in force measuring sensors.

➢ Potential use of robots with tactile sensing capabilities would be in assembly and inspection operations.

➢ In assembly, the robot could perform delicate part alignment and joining operations. In inspection, touch sensing would be used in gauging operations and dimensional measuring activities.

- Proximity sensors are used to sense when one object is close to another object. On a robot, the proximity sensors would be located n or near the end effectors.
- This sensing capability can be engineered by means of optical proximity devices, eddy-current proximity detectors, magnetic field sensors, or other devices.
- In robotics, proximity sensors might be used to indicate the presence or absence of a work part or other object. They could also be helpful in preventing injury to the robots human co-workers in the factory

Voice sensors

- Another area of robotics research is voice sensing or voice programming. Voice programming can be defined as the oral communication of commands to the robot or other machine.
- The robot controller is equipped with a speech recognition system which analyzes the voice input and compares it with a set of stored word patterns when a match is found between the input and the stored vocabulary word the robot performs some actions which corresponds to the word. Voice sensors could be useful in robot programming to speed up the programming procedure just as it does in NC programming.
- It would also be beneficial in especially in hazardous working environments for performing unique operations such as maintenance and repair work.
- The robot could be placed in hazardous environment and remotely commanded to perform the repair chores by means of step by step instructions.

Actuators used in the robotics

1. Hydraulic drive: gives a robot great speed and strength. These systems can be designed to actuate linear or rotational joints. The main disadvantage of a hydraulic system is that it occupies floor space in addition to that required by the robot.

2. Electric drive: compared with a hydraulic system, an electric system provides a robot with less speed and strength. Accordingly, electric drive systems are adopted for smaller robots. However, robots supported by electric drive systems are more accurate, exhibit better repeatability, and are cleaner to use.

3. Pneumatic drive: are generally used for smaller robots. These robots, with fewer degrees of freedom, carry out simple pick-and-place material handling operations.

The comparative advantages and dis advantages of these three drive system are listed below

Parameter	Hydraulic	Pneumatic System	Electric System
Accuracy	Medium	Low	High
Leakage	High	low	Nil
Load carrying capacity	High	Low	Medium
Maintenance	High	Low	Medium
Range of Speed	High	Low	Medium
Noise	High	Low	Medium
Work Area	Un Clean	Clean	Clean
Floor Space	More	Medium	Less
Repeatability	Medium	Low	High

ROBOT APPLICATIONS

Need to replace human labor by robots:

- Work environment hazardous for human beings
- Repetitive tasks
- Boring and unpleasant tasks
- Multi shift operations
- Infrequent changeovers
- Performing at a steady pace
- Operating for long hours without rest

- Responding in automated operations
- Minimizing variation

Industrial Robot Applications can be divided into:

Material-handling applications:

➢ Involve the movement of material or parts from one location to another.

- It includes part placement, palletizing and/or depalletizing, machine loading and unloading.

➢ Processing Operations:

- Requires the robot to manipulate a special process tool as the end effectors.
- The application include spot welding, arc welding, riveting, spray painting, machining, metal cutting, debarring, polishing.

➢ Assembly Applications:

- Involve part-handling manipulations of a special tools and other automatic tasks and operations

➢ Inspection Operations:

- Require the robot to position a work part to an inspection device.
- Involve the robot to manipulate a device or sensor to perform the
- inspection

➢ **Material Handling Applications**

- This category includes the following:
- Part Placement
- Palletizing and/or depalletizing

➢ Part Placement:

- The basic operation in this category is the relatively simple pick-and-placeoperation.
- This application needs a low-technology robot of t cylindrical
- Coordinate type.
- Only two, three, or four joints are required for most of the applications.
- Pneumatically powered robots are often utilized.

- **Processing operations:**

- Robot performs a processing procedure on the part.
- The robot is equipped with some type of process tooling as its end effectors.
- Manipulates the tooling relative to the working part during the cycle.
- Industrial robot applications in the processing operations include:
- Spot welding
- Continuous arc welding

Chapter 5

Manufacturing Process and Additive Manufacturing

5.1 INTRODUCTION TO CASTING

Metal Casting is one of the oldest materials shaping methods known. Casting means pouring molten metal into a mold with a cavity of the shape to be made, and allowing it to solidify. When solidified, the desired metal object is taken out from the mold either by breaking the mold or taking the mold apart. The solidified object is called the casting. By this process, intricate parts can be given strength and rigidity frequently not obtainable by any other manufacturing process. The mold, into which the metal is poured, is made of some heat resisting material. Sand is most often used as it resists the high temperature of the molten metal. Permanent molds of metal can also be used to cast products.

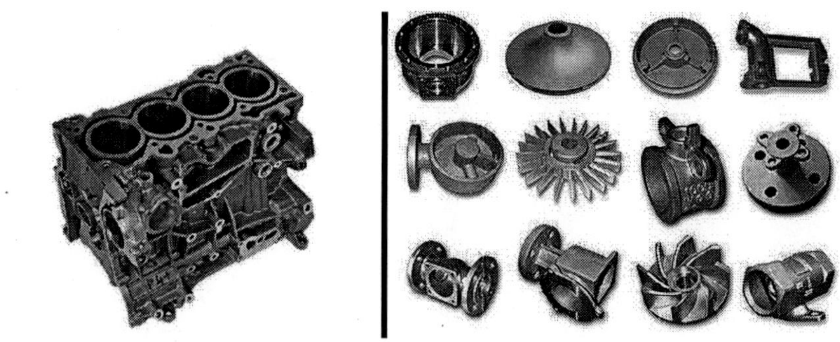

Figure 5.1 Casted components

In this process, components are produced by pouring molten metal into a contoured cavity followed by cooling to a solid mass.

1. The cold solid mass represents the configuration of the cavity and is the required shape of the component.

2. The components thus produced are called as casting.

3. The cavity compound to shape of the component is called mould.

4. The mould can be made of refractory material or metal.

5. The mould made of refractory material is called sand moulds and that of made from metal is called metal moulds or dies.

6. Cooling of liquid metal to solid metal is termed as phase transformation. The place where this activity is carried out is referred to as foundry.

7. In short, casting process involves shaping of the metal by using a mould cavity and hot metal.

8. In this process, the final shape is realized without using any other mechanism unlike in other processes, except the conversion of liquid metal to solid metal.

TERMINOLOGIES IN CASTING

a. Mould box: It is usually a metallic frame used for making for holding a sand mould. The mould box has two parts. The upper part is called "Cope" and the lower part is called "Drag".

b. Parting line: It is the zone of separation between a cope & drag position of the mould in sand casting.

c. Sprue/Runner: It is a vertical passage through which the molten metal will enter the gate.

d. Pouring basin: The enlarge position of the sprue with its top into which the molten metal is poured.

e. Gate: It is a short passageway which carries the molten metal from the runner sprue into the mould cavity.

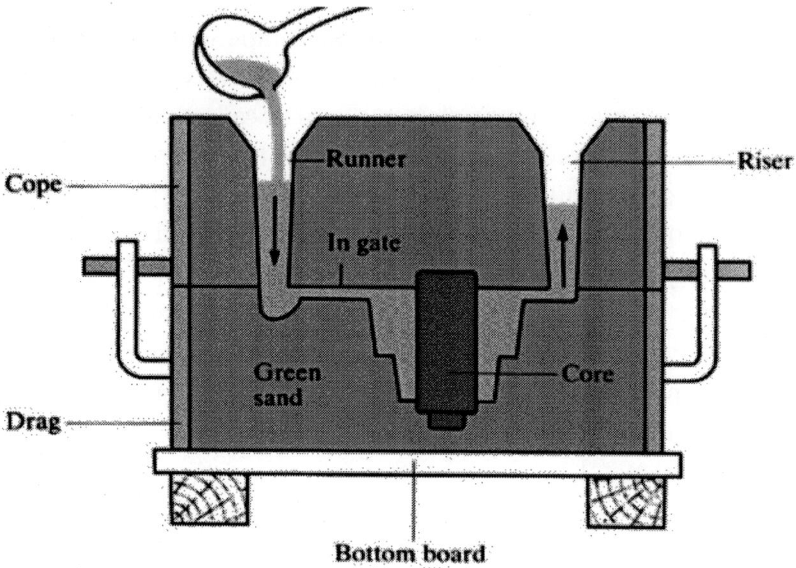

Figure 5.2 mold box and terminologies

f. Riser: A riser or a feed head is a vertical passage that stores the molten metal and supplies the same to the casting as it solidifies.

g. Mould cavity: The space in a mould that is filled with molten metal to form the casting upon solidification.

h. Core: A core is performed by using a mass of sand placed in the mould cavity to form hollow cavity in the castings.

i. Core print: It is a projection attached to the pattern to help for support and correct location of core in the mould cavity.

STEPS INVOLVED IN MAKING A CASTING

1. Pattern
2. Mould preparation
3. Core making
4. Melting and pouring
5. Cleaning and Inspection

1. Pattern

The pattern is the principal tool during the casting process. It is the replica of the object to be made by the casting process, with some modifications. The main modifications are the addition of pattern allowances, and the provision of core prints. If the casting is to be hollow, additional patterns called cores are used to create these cavities in the finished product. The quality of the casting produced depends upon the material of the pattern, its design, and construction. The costs of the pattern and the related equipment are reflected in the cost of the casting. The use of an expensive pattern is justified when the quantity of castings required is substantial. Some materials used for making patterns are: wood, metals and alloys, plastic, plaster of Paris, plastic and rubbers, wax, and resins.

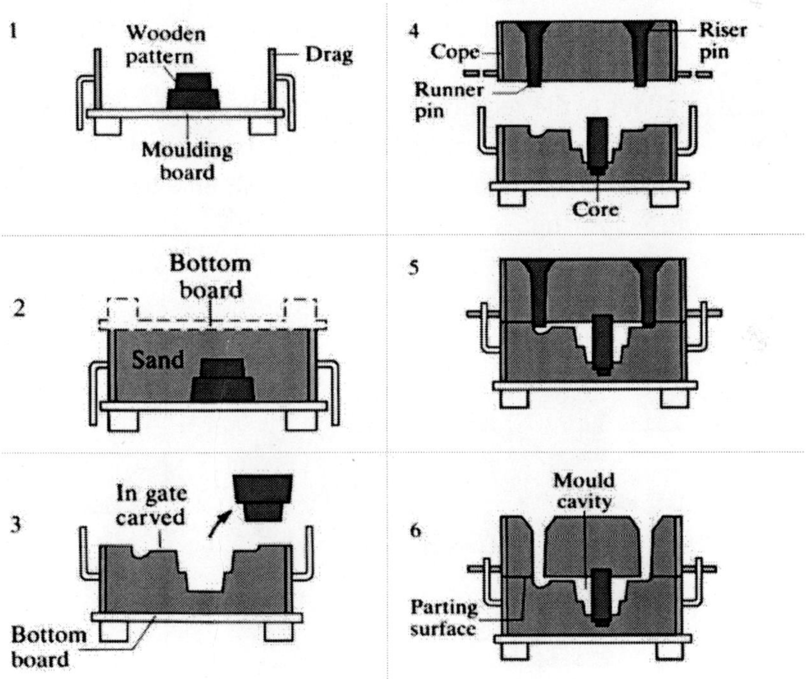

Figure 5.3 Steps in Sand moulding process

2. Mould preparation

Molding consists of all operations necessary to prepare a mold for receiving molten metal. It involves for making a cavity by packing sand around a pattern enclosed in a supporting metallic frame. When the pattern is removed from the mould an exact shaped cavity remains into which the molten metal is poured. Gating and risering are provided at suitable locations in the mould.

3. Core making

Castings are often required to have holes, recesses, etc. of various sizes and shapes. These impressions can be obtained by using cores. So where coring is required, provision should be made to support the core inside the mold cavity. Core prints are used to serve this purpose. The core print is an added projection on the pattern and it forms a seat in the mold on which the sand core rests during pouring of the mold.

4. Melting and pouring

Metals or alloys of the required composition are melted in a furnace and poured into the mould cavity. Many factors like temperature of molten metal, pouring time, turbulence, etc should be considered while making & pouring.

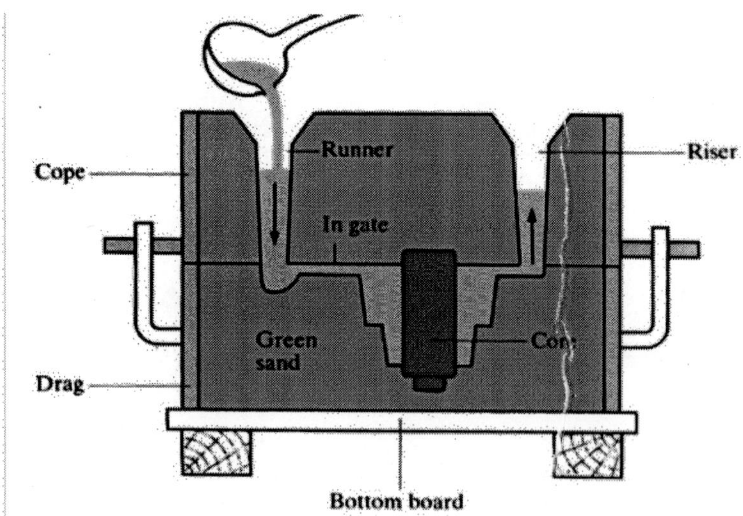

Figure 5.4 Sand Moulding

5. Cleaning and Inspection

After the molten metal is solidified and cooled, the rough casting is removed from the mould, cleaned and dressed (removing cores, adhered sand particles, gating, riser, fins, blisters, etc. from the casting surface) and then sent for inspection to check for dimensions or defects like blow holes, crakes, etc.

COMPONENTS PRODUCED BY CASTING PROCESS

To list the components produced by casting is an endless process, a few major components produce by casting are given below:

a. **Automotive sector**: A few parts like brake drum, cylinder, cylinder linings, pistons, engine blocks, universal joints, rocker arms, brackets, etc.

b. **Aircraft:** Turbine blades, casing, etc.

c. **Marine:** Propeller blades

d. **Machining:** Cutting tools, machine beds, wheels and pulleys, blocks, table for supports

e. **Agriculture & railroad equipment's:** Pumps and compressor, frames, valves, pipes and fittings for construction.

f. Camera frame, parts in washing machine, refrigerator & air-conditioners

g. Steel utensil & a wide variety of products.

Figure 5.5 Casted Metallic Parts

ADVANTAGES-DISADVANTAGES - APPLICATIONS OF CASTING

Advantages:

- Casting is the basic and versatile (flexible) manufacturing process.
- Large, hollow & intricate shapes can be easily cast.
- Casting provides freedom of design with respect to shape, size and quality of the product.
- Some metals that cannot be machined can be produced by casting to the required shape.
- Heavy objects can be produced only by casting process.
- Casting process is most suitable for mass production.
- Parts with close dimensional tolerance & ready to use can be produced by special casting methods.

Disadvantages:

- Casting process is an elaborate process and involves a greater number of operations.

- It requires large infrastructure like casting section, melting, pattern, core section, cleaning, finishing, and inspection.
- Very high investment is required.
- Casting process is not economical & viable for small scale operations.
- The manpower requirement is large.
- High care is required in handling chemicals and molten metal.
- Very difficult to cast thin sections.

Application of castings in different fields

- **Transport:** Automobile, aerospace, railways and shipping
- **Heavy Equipment:** Construction, farming and mining
- **Machine Tools:** Machining, casting, plastics molding, forging, extrusion and forming
- **Plant Machinery:** Chemical, petroleum, paper, sugar, textile, steel and thermal plants
- **Defense:** Vehicles, artillery, munitions, storage and supporting equipment
- **Electrical Equipment Machines:** Motors, generators, pumps and compressors
- **Hardware:** Plumbing industry pipes, joints, valves and fittings
- **Household:** Appliances, kitchen and gardening equipment, furniture and fittings
- **Art Objects:** Sculptures, idols, furniture, lamp stands and decorative items

CASTING DEFECTS

Casting process involves a number of variables, and a loss of control in any of these variables can cause defects under certain circumstances. Some of the common casting defects, their features and remedies to prevent such defects are discussed below.

(a) Shrinkage defect

Shrinkage is a void on the surface of the castings resulting from contraction or shrinkage of metal during solidification. Although a riser is used to overcome the shrinkage effect, in some cases it fails to feed the molten metal efficiently to the casting as it solidifies.

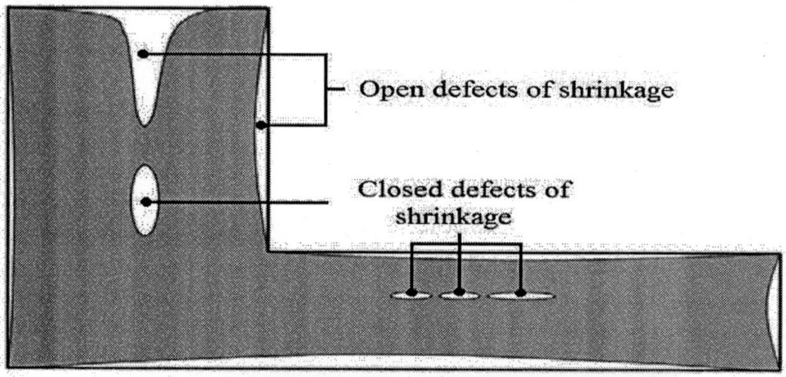

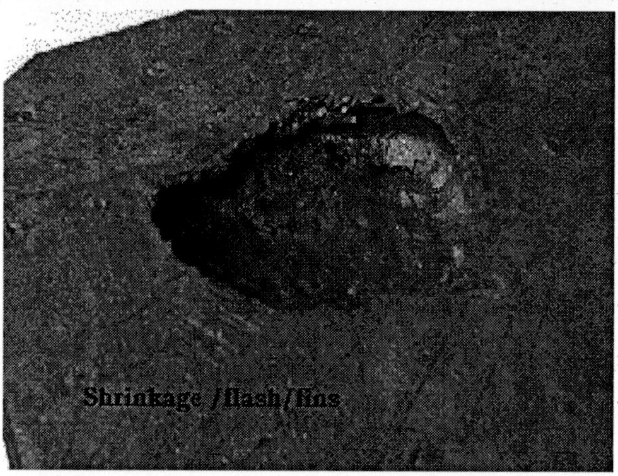

Figure 5.6 Shrinkage defect

Remedies

- Use large sprue and riser to promote directional solidification.
- Locate risers and gating systems in correct positions.
- Gates to be cut as wide as possible.

(b) Porosity defect (Blow hole and Pin hole)

Molten metal absorbs gases from various sources such as fluxes, moisture in sand, binders, additives and normal atmospheric gases like oxygen and nitrogen. If these gases are not allowed to escape, they get entrapped in the mould cavity forming small balloon shaped voids or cavities leading to porosity defect in castings. Two types of gas related defects occur in castings. They are blow hole and pin hole defect.

Blow holes occur below the surface of the castings and are not visible from the outside surface. Refer figure (a). On the other hand, pin holes are small gas cavities, many in number at or slightly below the surface of the casting. Refer figure (b).

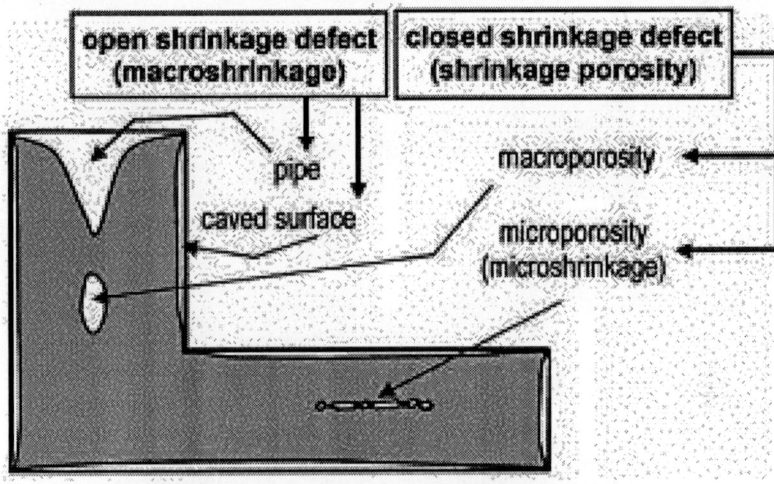

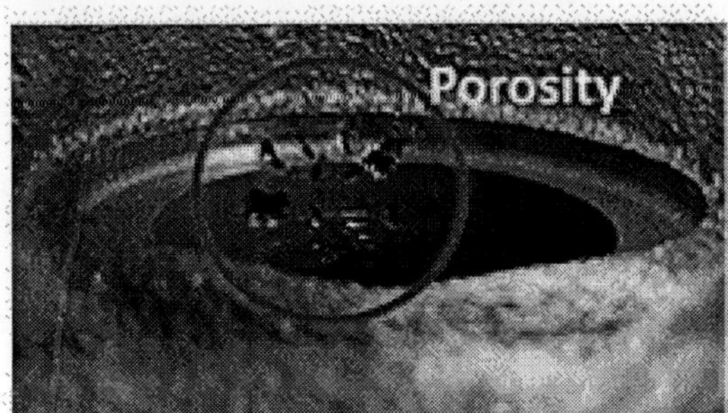

Figure 5.7 Porosity

Remedies

- Avoid excess ramming of mould.
- Provide proper vent holes.
- Avoid use of excess carbonaceous or other organic material in the sand/core binders, because these materials react with the molten metal producing large amount of gases.

(c) Misrun

Misrun occur when the mould cavity is not fully filled with molten metal. In other words, it is a defect wherein a casting solidifies before the molten metal completely fills the cavity.

Remedies

- Fluidity of metal should be maintained suitably.
- Pouring rate and time should be controlled.
- Thin sections should be suitable designed.

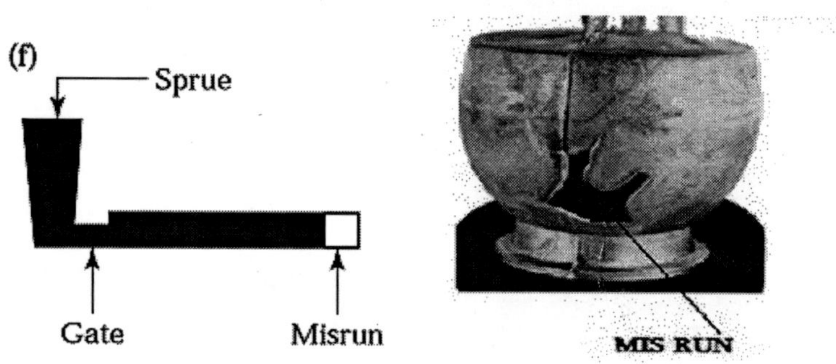

Figure 5.8 Misrun

(d) Penetration

When fluidity of liquid metal is high, it may penetrate the sand mould/core (into the voids between the sand particles). A fused aggregate of metal and sand appears on the surface of the casting leading to defect.

Remedies

- Sand should be properly rammed.
- Moulding sand/core sand should not be too coarse to promote metal penetration.

- Control proper metal temperature. Fluidity of molten metal should be maintained suitably.

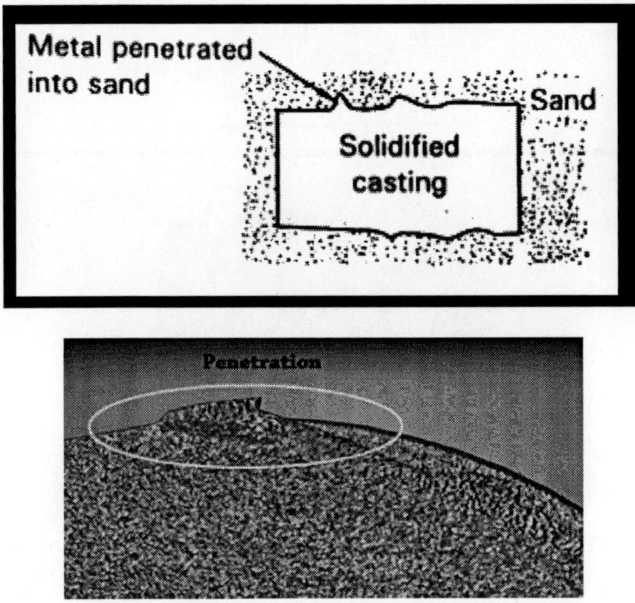

Figure 5.9 Penetration

(e) Mould shift (Mismatch)

It is a step produced in the cast product along the parting line due to the sidewise relative displacement of cope and drag box.

Remedies

- Proper alignment of cope and drag box.
- Proper handling of assembled cope and drag box during operations.

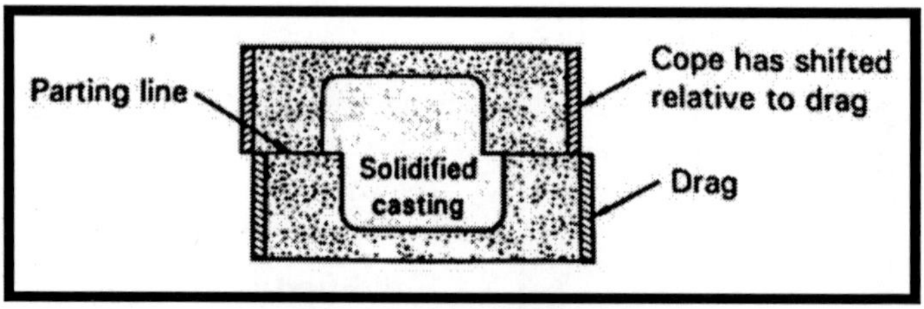

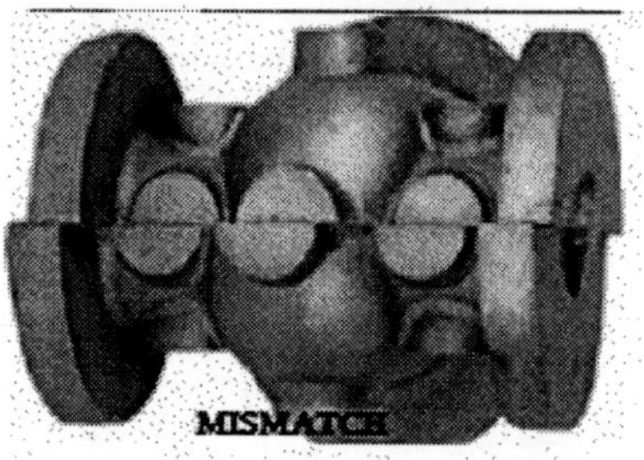

Figure 5.10 Mismatch

(f) Cold shut

Two portions of metal flow together, but lack of fusion due to premature freezing results in a defect known as cold shut.

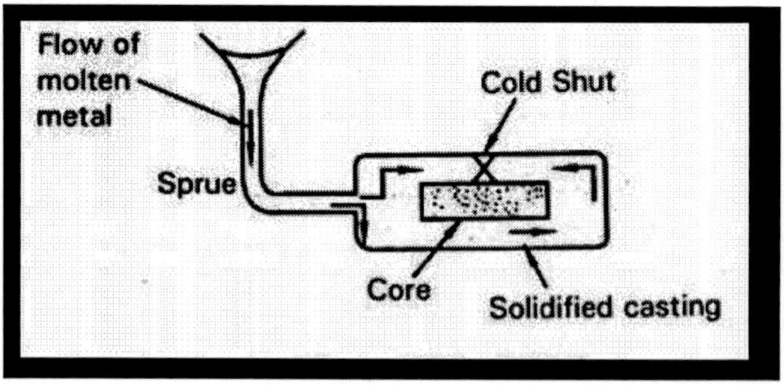

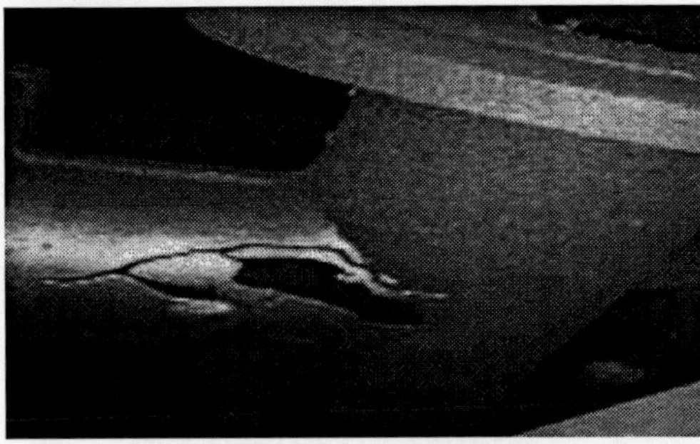

Figure 5.11 Cold Shut

Remedies

- Place gates and risers at proper locations.
- Metal fluidity should be high.

(g) Hot tears

A hot tear is an internal or external ragged discontinuity formed in the casting due to the pulling action of the metal just after it has solidified.

Remedies

- Provide adequate fillets at sharp corners while designing the component.
- Proper metallurgical and pouring temperature to be maintained.

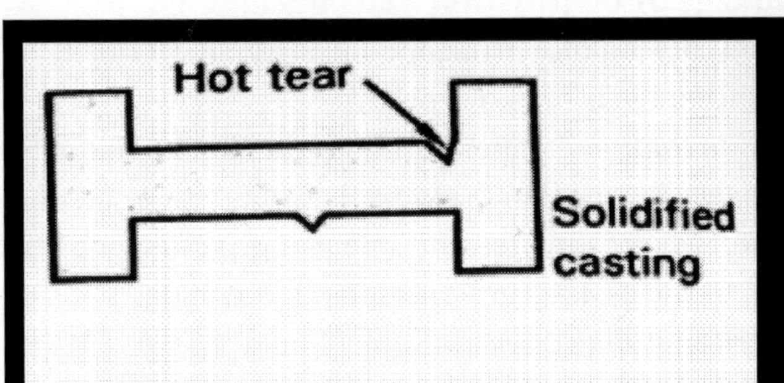

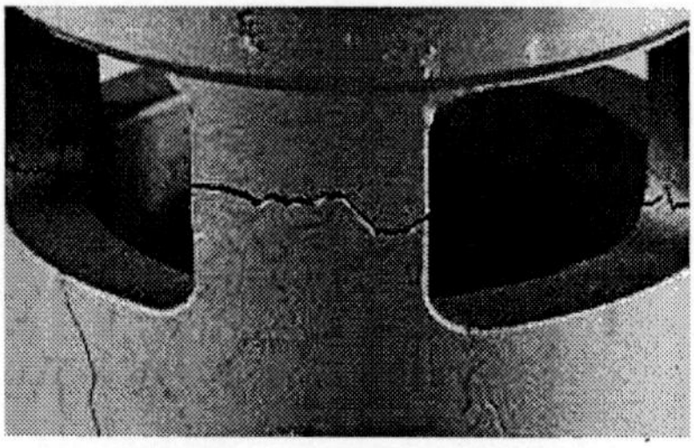

Figure 5.12 Hot Tear

5.2 PATTERN ALLOWANCES

Pattern allowance is a vital feature as it affects the dimensional characteristics of the casting. Thus, when the pattern is produced, certain allowances must be given on the sizes specified in the finished component drawing so that a casting with the particular specification can be made. The selection of correct allowances greatly helps to reduce machining costs and avoid rejections. The allowances usually considered on patterns and core boxes are as follows:

1. Shrinkage or contraction allowance
2. Draft or taper allowance
3. Machining or finish allowance
4. Distortion or camber allowance
5. Rapping allowance

1. Shrinkage or Contraction Allowance

All most all cast metals shrink or contract volumetrically on cooling. The metal shrinkage is of two types:

i. *Liquid Shrinkage:* it refers to the reduction in volume when the metal changes from liquid state to solid state at the solidus temperature. To account for this shrinkage; riser, which feed the liquid metal to the casting, are provided in the mold.

ii. *Solid Shrinkage:* it refers to the reduction in volume caused when metal loses temperature in solid state. To account for this, shrinkage allowance is provided on the patterns.

The rate of contraction with temperature is dependent on the material. For example steel contracts to a higher degree compared to aluminum. To compensate the solid shrinkage, a shrink rule must be used in laying out the measurements for the pattern.

2. Draft or Taper Allowance

The taper provided by the pattern maker on all vertical surfaces of the pattern so that it can be removed from the sand without tearing away the sides of the sand mold and without excessive rapping by the molder is known as draft. Figure 7.6 is an illustration of a pattern having proper draft allowance. Here, the moment the pattern lifting commences, all of its surfaces are well away from the sand surface. Thus, the pattern can be removed without damaging the mold cavity.

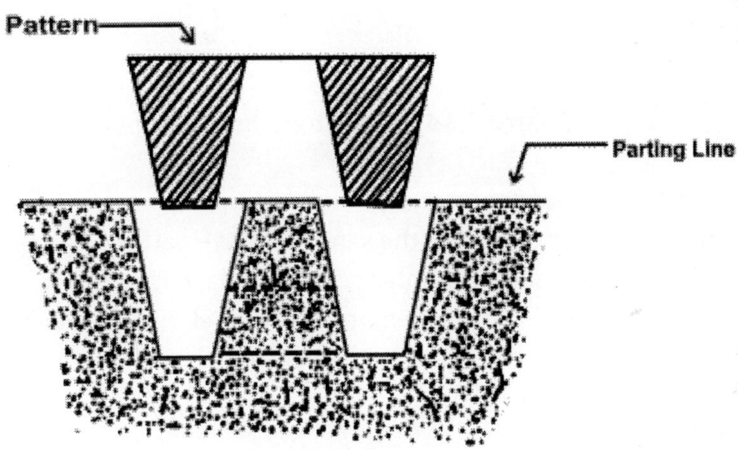

Figure 5.13 Pattern having draft on vertical edges

3. Machining or Finish Allowance

The finish and accuracy achieved in sand casting are generally poor and therefore when the casting is functionally required to be of good surface finish or dimensionally accurate, it is generally achieved by subsequent machining. Machining or finish allowances are therefore added in the pattern dimension.

4. Distortion or Camber Allowance

Sometimes castings get distorted, during solidification, due to their typical shape. For example, if the casting has the form of the letter U, V, T, or L etc. it will tend to contract at the closed end causing the vertical legs to look slightly inclined. This can be prevented by making the legs of the U, V, T, or L shaped pattern converge slightly (inward) so that the casting after distortion will have its sides vertical.

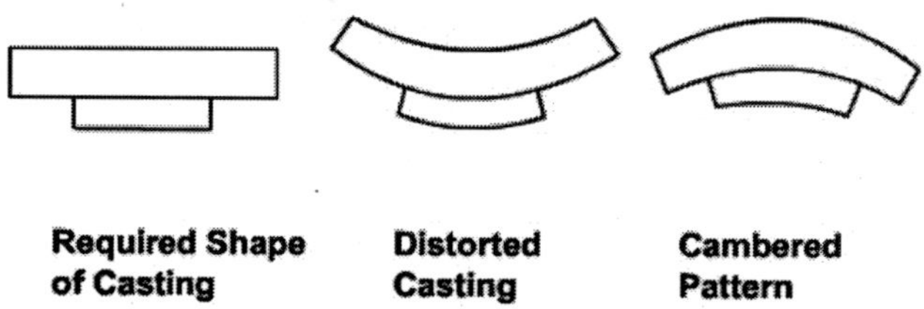

Figure 5.14 Distortions in casting

5. **Rapping Allowance**

Before the withdrawal from the sand mold, the pattern is rapped all around the vertical faces to enlarge the mold cavity slightly, which facilitate its removal. Since it enlarges the final casting made, it is desirable that the original pattern dimension should be reduced to account for this increase.

Problem 1:

A job shown in the Figure is to be made of steel by casting process. The mould for this job is made from a wooden pattern. Determine the dimensions of the wooden pattern. Assume machining allowance of 2 mm on each side, shrinkage allowance of 2% and a taper allowance of 1 degree.

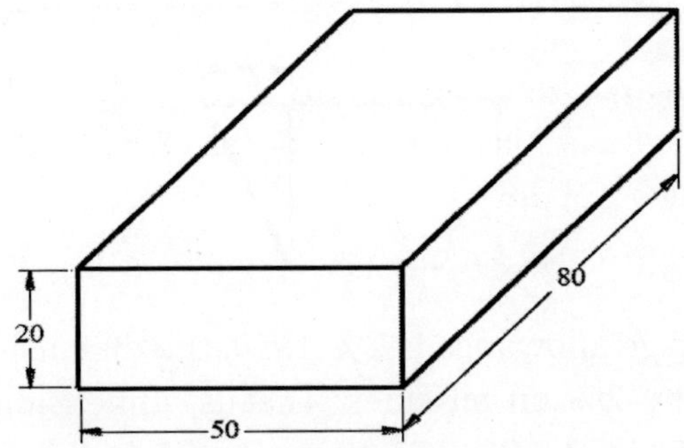

Solution:

It is given that machining allowance of 2 mm on each side is to be given. Thus, each side is increased by 2 mm resulting in the basic dimension of the pattern. The required casting is shown with dotted lines.

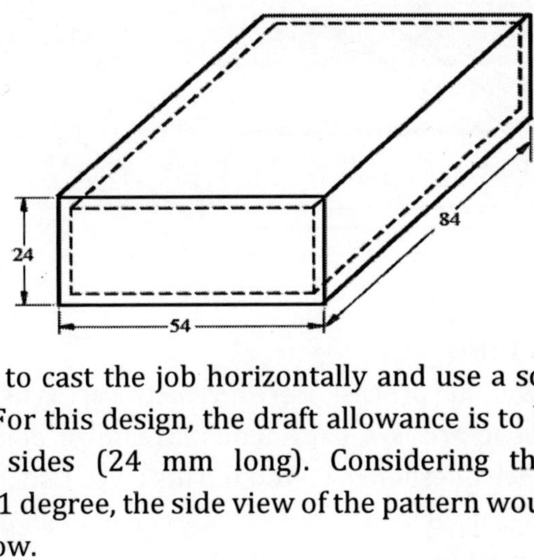

We decide to cast the job horizontally and use a solid pattern for this casting. For this design, the draft allowance is to be provided on the vertical sides (24 mm long). Considering the given taper allowance of 1 degree, the side view of the pattern would be as shown in Figure below.

The taper allowance value x is calculated from the geometry of the Figure below as

X = 24 tan (1) = 0.419 mm.

Thus, the top surface dimension is increased to provide for draft allowance from 54 x 84 mm to 54.838 x 84.838 mm.

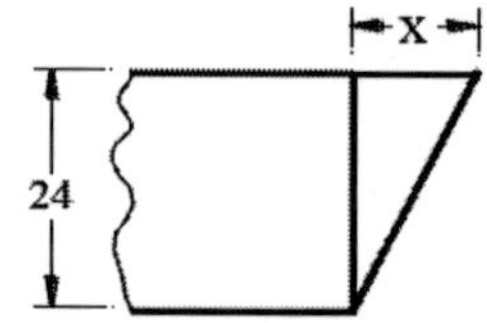

Given shrinkage allowance is 2%. Now, the dimensions of pattern are increased by 2% on all sides. That is, dimension 54 mm will become 54 + (54 x 2)/100 = 55.08 mm or 55.1 mm. The dimension 54.838 will become 54.838 + (54.838 x 2)/100 = 55.9 mm

Similarly, all other dimensions are calculated and the final dimensions of the pattern are shown in Figure below.

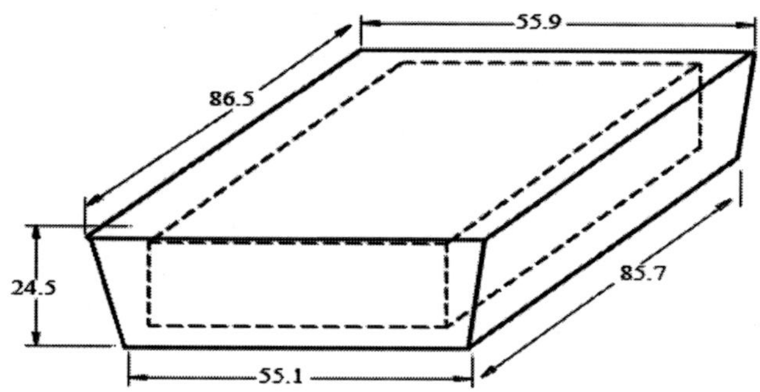

5.3 ARC WELDING

It is a fusion welding process. Here an electric arc is used as the heat source. An electric arc is a continuous stream of electron flowing between two electrodes into a medium. This gives rise to intense heat generation. This is used to melt the surface and deposit the metal from the electrode resulting in a joint on cooling.

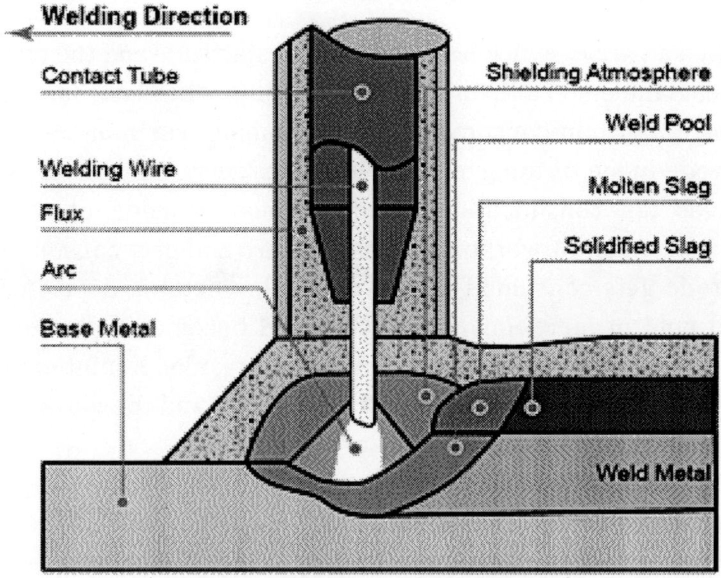

Figure 5.14 Arc Welding Zone

7.2.2 PRINCIPLE OF ARC WELDING

The source of heat for arc welding process is an 'electric arc' generated between two electrically conducting materials. One of the workpiece materials called electrode is connected to one pole of the electric circuit, while the other workpiece which forms the second conducting material is connected to the other pole of the circuit.

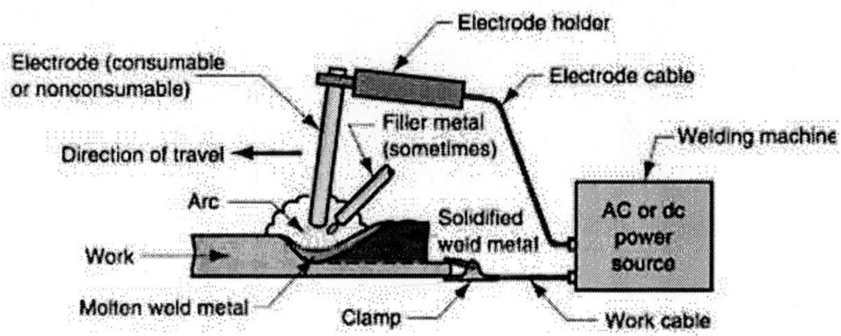

Figure 5.15 Components of Arc welding process

When an arc is struck between a metal electrode and the work, the arc heats the metal and forms a pool of liquid. The electrode material can be either non-consumable or consumable. The non-consumable electrode made of tungsten, graphite, etc., serve only strike the arc, whereas the consumable electrode which is made of the same material as that of workpiece strike the arc and gets consumed. The electrode gets consumed as and when the metal is deposited. The liquid pool of metal mixes the work and becomes a part of it. On cooling, solid metal of homogeneous bond results. A joint is formed. The distance between the tip of the electrode and the work through which arc is formed is termed as 'arc length'.

WELDING ELECTRODE

An electrode is a metal wire that is coated. It is made out of materials with a similar composition to the metal being welded. There are a variety of factors that go into choosing the right electrode for each project. In summary:

- SMAW or stick electrodes are consumable, meaning they become part of the weld and are also referred to as a filler electrode or welding rod.
- TIG tungsten electrodes are non-consumable as they do not melt and become part of the weld, requiring the use of a welding rod.
- TIG filler rods are an optional filler material used to fuse two pieces of stock together as a composite.
- The MIG welding electrode is a continuously fed wire referred to as MIG wire.

Electrode selection is critical to ease of cleanup, weld strength, bead quality, and minimizing any spatter. Electrodes need to be stored in a moisture-free environment and carefully removed from any package (follow the directions to avoid damage).

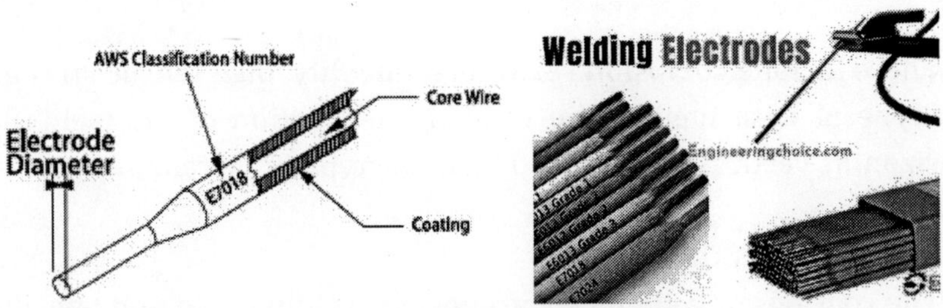

Figure 5.16 Welding rods

When molten metal is exposed to air, it absorbs oxygen and nitrogen and becomes brittle or is otherwise adversely affected. A slag cover is needed to protect molten or to solidify weld metal from the atmosphere. This cover can be obtained from the electrode coating.

The composition of the welding electrode coating determines its usability, the composition of the deposited weld metal, and the electrode specification. The formulation of welding electrode coatings is based on well-established principles of metallurgy, chemistry, and physics.

The coating protects the metal from damage, stabilizes the arc, and improves the weld in other ways, which include:

1. Smooth weld metal surface with even edges
2. Minimum spatter adjacent to the weld
3. A stable welding arc
4. Penetration control
5. Easier slag removal
6. Improved deposition rate

The metal-arc electrodes may be grouped and classified as bare or thinly coated electrodes and shielded arc or heavy coated electrodes. The covered electrode is the most popular type of filler metal used in arc welding. The composition of the electrode covering determines the usability of the electrode, the composition of the deposited weld metal, and the specification of the electrode.

The type of electrode used depends on the specific properties required in the weld deposited.

These include corrosion resistance, ductility, high tensile strength, the type of base metal to be welded, the position of the weld (flat, horizontal, vertical, or overhead), and the type of current and polarity required.

ELECTRODE COATINGS

The coatings of welding electrodes for welding mild and low alloy steels may have from 6 to 12 ingredients, which include:

- **Cellulose** – to provide a gaseous shield with a reducing agent in which the disintegration of cellulose produces the gas shield surrounding the arc
- **Metal carbonates** – to adjust the basicity of the slag and to provide a reducing atmosphere
- **Titanium dioxide** – to help form a highly fluid but quick-freezing slag and to provide ionization for the arc
- **Ferromanganese & ferrosilicon** – to help deoxidize the molten weld metal and supplement the manganese content and silicon content of the deposited weld metal.
- **Clays & gums** – to provide elasticity for extruding the plastic coating material and to help provide strength to the coating
- **Calcium fluoride** – to provide shielding gas to protect the arc, adjust the basicity of the slag, and provide fluidity and solubility of the metal oxides
- **Mineral silicates** – to provide slag and give strength to the electrode covering
- **Alloying metals including nickel, molybdenum, and chromium** – to provide alloy content to the deposited weld metal
- **Iron or manganese oxide** – to adjust the slag's fluidity and properties and help stabilize the arc.
- **Iron powder** – to increase productivity by providing extra metal to be deposited in the weld.

Advantages of Arc Welding:

1. It is suitable for high-speed welds.
2. It can work on AC or DC.
3. Portability as a result of the simple equipment.
4. It is a fast welding process when compared to others.
5. It offers strong joints.
6. Produces very little distortion.
7. It has the ability to weld on porous and dirty metal.
8. Equipment used in arc welding is inexpensive.
9. Its operation can be performed during wind or rain.

Disadvantages of Arc Welding:

1. Cannot be used for reactive metal like Aluminium or Titanium.
2. Not suitable for welding thin metals.
3. It is hazardous when not performed under the safety and security guidelines.
4. It is a difficult task to dismantle the joined material through welding.
5. Requires skilled labor and electric supply.

TUNGSTEN INERT GAS WELDING (TIG)

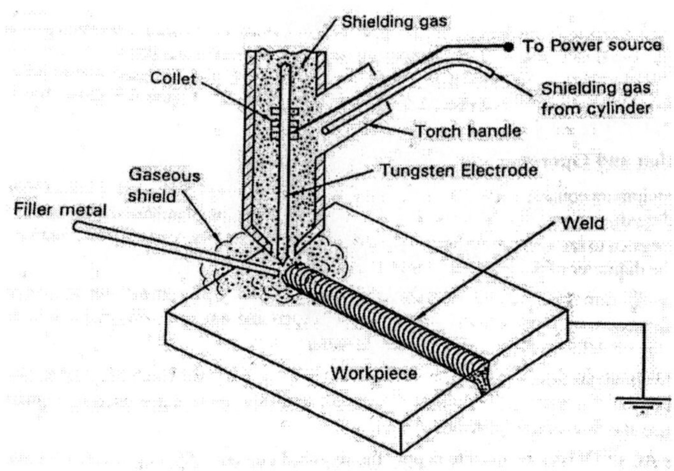

5.17 TIG welding process

An arc is struck between the non-consumable tungsten electrode and the workpiece in the presence of an inert gas such as helium or argon, which is passed near the arc thereby creates a shield. This is referred to as Tungsten Inert Gas Welding (TIG). A filler rod is used to supply molten metal to the joint. Fig. shows the TIG welding. Pressure of the gas varies between 0.1 to 0.15 N/mm^2. Mainly DC supply is used.

Advantages:

- Suitable for thin metals.
- Clear visibility of the arc provides the operator to have greater control over the weld.
- Strong and high quality of joints are obtained.
- No flux is used, no slag formation, results in clean weld joint.

Disadvantages:

- TIG is most difficult process compared to all the other welding process.
- Skilled operator is required.
- Process is slower.
- Not suitable for thick metals.

APPLICATIONS

TIG welding is most commonly used to weld stainless steel and non-ferrous materials such as aluminium and magnesium. Aerospace industry is one of the primary users, refrigerators, air conditioners and chemical plants are other users.

Figure 5.18 TIG Welded Parts

METAL INERT GAS WELDING (MIG)

Metal inert gas welding, in which the workpiece is joined by the heat obtained from an electric arc struck between a metal consumable electrode (uncoated) and the workpiece in the presence of an inert gas, which is a mixture of argon & carbon dioxide. The consumable metal electrode itself will supply the required metal. No filler is required.

Advantages:

- MIG welding is fast and economical.
- Weld deposition rate is high due to continuous wire feed.
- No flux is used, no slag formation, clean weld is obtained.
- Process can be automated.

Disadvantages:

- Equipment is costlier.
- Dross & porosity are the most prevalent quality problem.

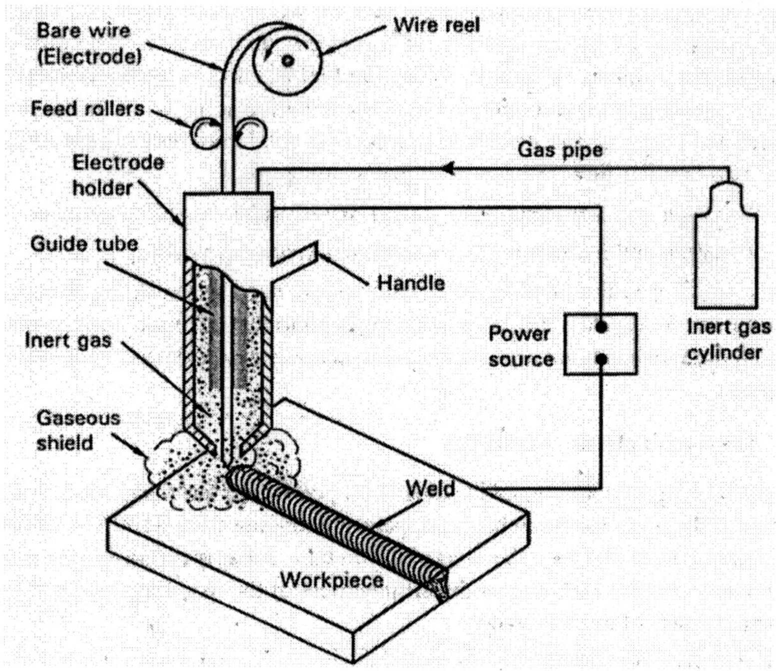

Figure 5.19 MIG welding Process

Some of the applications of MIG welding:

Used for most types of sheet metal welding, Fabrication of pressure vessels and steel structures, Automotive industry and home improvement industry.

Questions for Review

1. With a neat sketch explain the basic principle of electric arc welding. Label the sketch showing the details of the equipment used.
2. Explain the functions of shielding gases in arc welding. Name some shielding gases used in arc welding.
3. Explain the various functions that can be served by coating of the coated electrode.
4. State about benefits and limitations of i) TIG welding ii) MIG welding.
5. Explain the function of flux in welding.

6. Explain the principle of Arc welding. What are the various weld positions in arc welding? Enumerate some defects due to Arc welding.
7. Compare MIG and TIG welding in respect to their principle of working and field of application.
8. Demonstrate metal inert Gas arc welding process with a neat sketch.
9. Compare and contrast the difference between welding and brazing.

5.4 BRAZING

INTRODUCTION

Brazing is a method of joining similar or dissimilar metals by means of a filler metal whose melting temperature is above 450°C, but below the melting point of the base metal.

The filler metal called speller is a non-ferrous metal or alloy. Copper and copper alloys, silver and silver alloys, and aluminum alloys are the most commonly used filler metals.

The flow of molten filler material into the gap between the two workpieces is driven by the capillary force. The filler material cools down and solidifies forming a strong joint.

The base metals are not fused in the process.

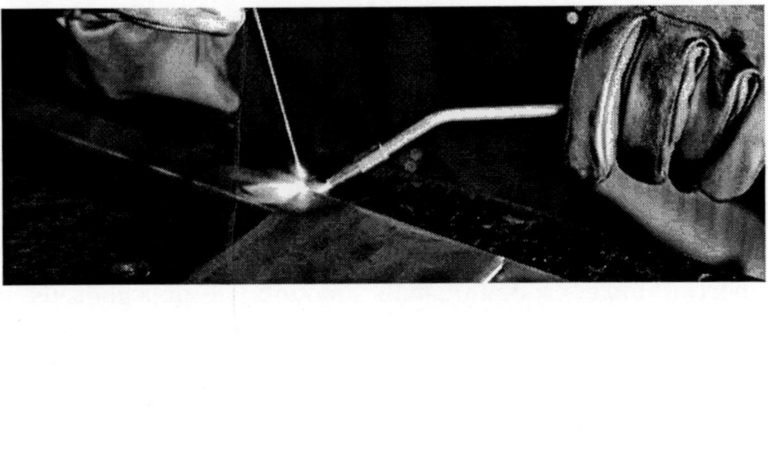

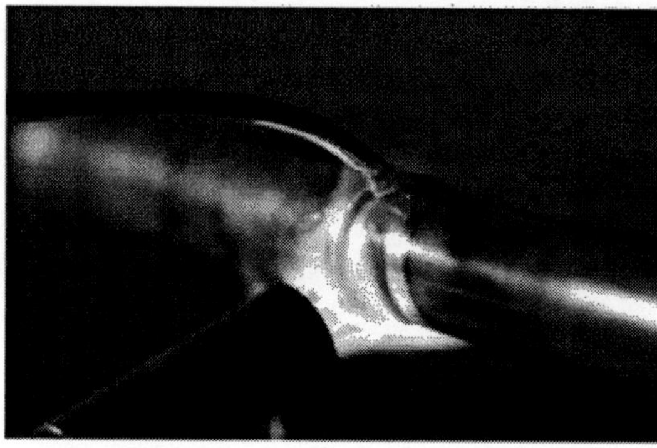

Figure 5.20 Brazing Process

FLUX USED IN BRAZING

Flux performs its usual function as in soldering, i.e., it melts during the preheating stage and spreads over the joint area, wetting it and protecting the surface from oxidation. It also cleans the surface, dissolving the metal oxides.

The flux used in brazing is available in powder, liquid, and paste form. One method of applying the flux in powdered form is to dip the heated end of the filler rod into the container of the powdered flux, and allowing the flux to stick to the filler rod. Another method is to heat the base metal slightly and sprinkle the powdered flux over the joint, allowing the flux to partly melt and stick to the base metal. Sometimes, it is desirable to mix powdered flux with clean water (distilled water) to form a paste. Flux in either the paste or liquid form can be applied with a brush to the joint. Better results occur when the filler metal is also given a coat. The most common type of flux used is borax, or a mixture of borax with other chemicals. Some of the commercial fluxes contain small amounts of phosphorous and halogen salts of iodine, bromine, fluorine, chlorine, or astatine. When a prepared flux is not available, a mixture of 12 parts of borax and 1-part boric acid may be used.

Torch brazing

Torch brazing is a brazing process in which the two metals are joined by the heat obtained with a gas flame, and by using a non-ferrous filler metal having a melting temperature of above 450°C, but below the melting temperature of the base metal.

In operation, the surfaces to be joined are cleaned thoroughly in order to remove dirt, grease and other oxides. After cleaning, flux is applied at the joint. The base metals are broadly heated by an oxy-acetylene welding torch. The filler metal is then placed at the joint and is heated with a carburizing flame. The filler metal melts and flows through the joint by capillary action. The workpiece is allowed to cool for some time. The joint is cleaned to remove flux residues in order to prevent corrosion.

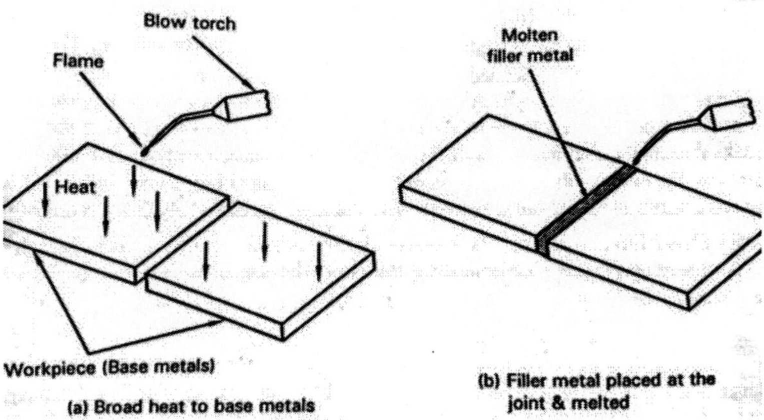

Figure 5.21 Torch Brazing

Advantages

- Much heat is not involved in the process. Hence, low thermal distortions and residual stresses in the joints.
- Easily automated process.
- Dissimilar parts can be joined.

Disadvantages

- Flux residues must be removed after brazing, otherwise which may cause corrosion.
- Large and thick sections cannot be brazed efficiently.

Applications

1. Automotive (e. g. , joining tubes and pipes)
2. Electrical equipment (e. g. , joining wires and cables)
3. Cutting tools (e. g. , brazing cemented carbide inserts to shanks)
4. Jewelry
5. Chemical process industry
6. Plumbing and heating contractors join metal pipes and tubes by brazing
7. Repair and maintenance work

Chapter 6

ADDITIVE MANUFACTURING

6.1 INTRODUCTION

Additive manufacturing or direct digital manufacturing, is a process for making a physical object from a three-dimensional digital model, typically by laying down many successive thin layers of a material. The successive layering of materials constitutes the technique of additive manufacturing. Thus, the term *direct digital manufacturing* stems from the process of going from a digital blueprint of a product to a finished physical product. Manufacturers can use 3D printing to make prototypes of products before going for full production.

Additive Manufacturing (rapid prototyping, 3D printing) can be defined as the layer-by-layer fabrication of three-dimensional physical models directly from a computer – aided design (CAD) data.

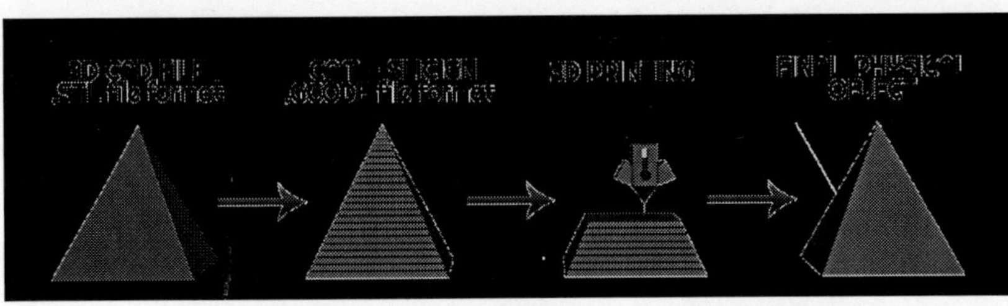

Figure 6.1 Steps in Additive manufacturing

The evolution of ADDITIVE MANUFACTURING (rapid prototyping, 3d printing, free form fabrication) has changed the face of direct, digital technologies for the rapid production of models, prototypes, patterns, and fit & functional parts, since its introduction, AM technology has changed design, engineering, and manufacturing processes within the aerospace, automotive, electrical & electronics, consumer industries, biomedical and dental devices & implants. Due to wide applications, rapid prototyping technology has become a revolutionary field in manufacturing.

6.2 STEPS IN ADDITIVE MANUFACTURING

1. The first step is to create a 3D model of the object to be printed. This model can be designed with computer-aided design (CAD) software or through reverse engineering techniques by for example using an object laser scanner.

2. The CAD file is then converted to a standard additive manufacturing file format - usually an STL file. This step is regarded as the simplest. Conversion to a STL file will tessellate shapes and the time needed to do this depends on PC power and model complexity. The file is then digitally sliced into layers.

3. The third stage requires the transfer of the STL file and setting up of the machine. In order to print economically i.e. maximise cost savings and reduce material waste, the positioning and sizing of the object on the build platform should be checked – usually multiple parts are printed in one go.

4. In the fourth step, the machine which is only controlled by a computer builds the model layer by layer. The layer thickness dictates the final quality and depends on the machine and process. For Fused Deposition Modelling a layer thickness of 0.254 mm is typical, for Stereolithography printing can result in a quality of 0.05 – 0.1 mm . The size of a model depends on the machine size.

5. After building the part and potentially applying cooling and curing periods, the model can be removed from the machine. Additional post processing can occur such as cleaning, polishing, painting and finishing of the surface to the desired standard. This may involve the use of other machines and tools.

ADVANTAGES AND DISADVANTAGES OF ADDITIVE MANUFACTURING

Advantages:

1) AM can print complex 3D geometries with internal features without any tooling

2) Reduced waste compared to machining

3) Part can be printed directly from the 3D model without the need for a drawing

4) Prototypes can be made quicker, allowing designers to check different iterations resulting in a quicker design cycle phase

5) No or less tooling for smaller batches compared to traditional machining

6) Production tooling can be printed

7) Different materials can be mixed during the printing process to create a unique alloy

8) Different sections of the part can be different variants of the same alloy

Disadvantages:

1) Because the technology is still in its infancy, the build process is slow and costly

2) High production costs because of the equipment cost

3) Various post-processing required depending on the type of additive manufacturing used

4) Small build volume compared to other manufacturing part size such as sand casting

5) Poor mechanical properties hence need post-processing

6) Poor surface finish and texture compared manufacturing processes like CNC and investment casting.

7) The strength of the parts is comparably weaker compared to manufacturing processes such as Die casting, Investment casting and CNC machining

CLASSIFICAION OF ADDITIVE MANUFACTURING

There are numerous ways to classify additive manufacturing technologies.

- According to baseline technology, like whether the process uses lasers, printer technology, extrusion technology, etc.
- According to processes and the type of raw material input. The problem with these classification methods is that some processes get

grouped together in what seems to be odd combinations (like selective laser sintering (SLS) being grouped together with 3d printing) or that some processes that may appear to produce similar results end up being separated (like stereolithographic and material jetting with photopolymers). It is probably inappropriate, therefore, to use a single classification approach.

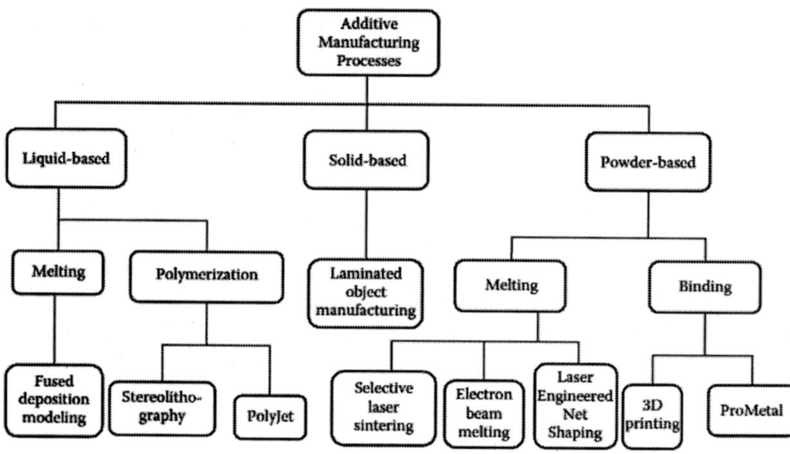

Figure 6.2 Classification of Additive Manufacturing Process

6.3 STEREO LITHOGRAPHY (SLA OR SL) SYSTEMS

Stereo Lithography (SL) is the first process ever developed in rapid prototyping field with the meaning of 3-dimensional printing. It is a liquid-based process that consists of the curing or solidification of a photosensitive polymer when an ultraviolet laser makes contact with the resin. The process begins with construction of a model using CAD software. The model is then translated to a STL file in which the pieces are cut into "slices," with each slice containing the information required for each layer. The thickness of each layer as well as its resolution depend on the equipment used.

Working Process of Stereo Lithography:

SLAs have four main parts a tank that can be filled with liquid plastic (photopolymer), a perforated platform that is lowered into the tank, an ultraviolet (UV) laser and a computer controlling the platform and the laser, as shown in figure 8.4.

In the initial step of the SLA process, a thin layer of photopolymer (usually between 0.05-0.15 mm) is exposed above the perforated platform. The UV laser hits the perforated platform, "painting" the pattern of the object being printed. The UV-curable liquid hardens instantly when the UV laser touches it, forming the first layer of the 3D-printed object. Once the initial layer of the object has hardened, the platform is lowered, exposing a new surface layer of liquid polymer. The laser again traces a cross section of the object being printed, which instantly bonds to the hardened section beneath it. This process is repeated again and again until the entire object has been formed and is fully submerged in the tank. The platform is then raised to expose a three-dimensional object. After it is rinsed with a liquid solvent to free it of excess resin, the object is baked in an ultraviolet oven to further cure the plastic.

Objects made using stereo lithography generally have smooth surfaces, but the quality of an object depends on the quality of the SLA machine used to print it. The amount of time it takes to create an object with stereo lithography also depends on the size of the machine used to print it. Small objects are usually produced with smaller machines and typically take between six twelve hours to print. Larger objects, which can be several meters in three dimensions, take days.

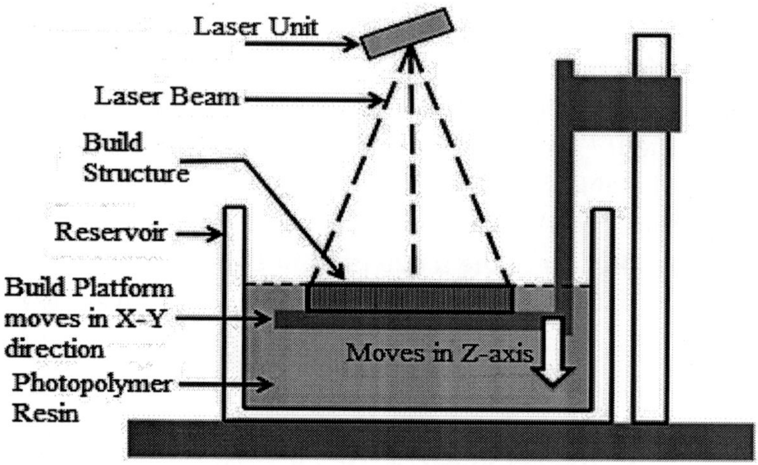

Figure 6.3 Stereo Lithography Process

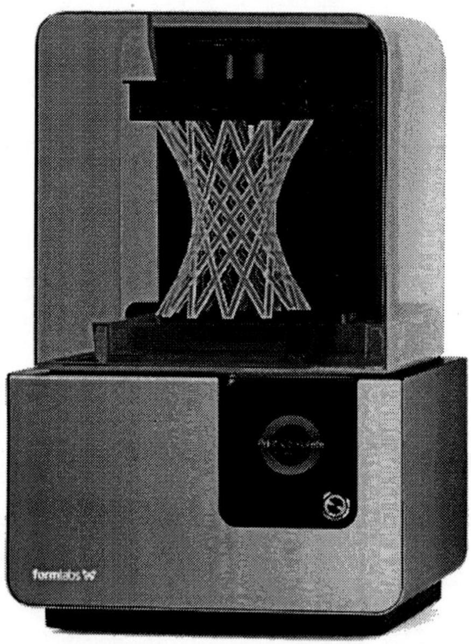

Fig 6.4 Stereo Lithography Apparatus

Advantages

1) The SLA can be used continuously and unattended round the clock.

2) The computerized process serves as a good user support.

3) The different SLA machines have built volumes ranging from small to large to suit the needs of different users.

4) The SLA has good accuracy and can thus be used for many application areas.

5) The SLA can obtain one of the best surface finishes amongst RP technologies.

6) There is a wide range of materials, from general-purpose materials to specialty materials for specific applications.

Disadvantages

1) Structures that have overhangs and undercuts must have supports that are designed and fabricated together with the main structure.

2) Post-processing includes removal of supports and other unwanted materials, which is tedious, time-consuming and can damage the model.

3) Post-curing may be needed to cure the object completely and ensure the integrity of the structure.

Applications of SLA:

The SLA technology provides manufacturers with cost justifiable methods for reducing time to market, lowering product development costs, gaining greater control of their design process and improving product design.

The range of applications includes:

1) Models for conceptualization, packaging and presentation.

2) Prototypes for design, analysis, verification and functional testing.

3) Parts for prototype tooling and low volume production tooling.

4) Patterns for investment casting, sand casting and moulding.

5) Tools for fixture and tooling design, and production tooling.

FUSED DEPOSITION MODELLING (FDM)

Like Selective Laser Sintering, FDM uses thermoplastics to print parts. The difference between these two is that FDM uses filaments, whereas SLS uses powders. Stereolithography differs from both as it uses thermosets.

Step 1: A geometric model of a conceptual design is created on CAD software which uses IGES or STL formatted files. It can then be imported into the workstation where it is processed through.

The Quick Slice and Support Work propriety software before loading to FDM systems. The CAD file is sliced into horizontal layers after the part is oriented for the optimum build position, and any necessary support structures are automatically detected and generated. The slice thickness can be set manually to anywhere between 0.172 to 0.356 mm (0.005 to 0.014 in) depending on the needs of the models.

Step 2: The nozzle is heated to melt the plastic filament and is mounted to a mechanical stage which can be moved in both horizontal directions. As the nozzle is moved over the table in the required geometry, it deposits a thin bead of extruded plastic to form each layer and create a two-dimensional cross section of the model. The plastic hardens immediately after being squirted from the nozzle and bonds to the layer below. The platform then descends where the next layer is extruded upon the previous. This continues until the model is completed. The entire system is contained within a chamber which is held at a temperature just below the melting point of the plastic.

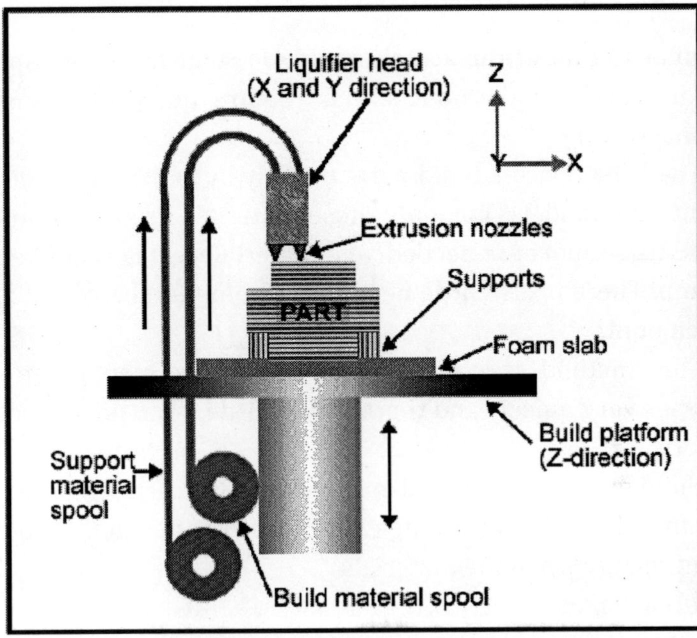

Figure 6.5 Fused Deposition Modeling

Step 3: Once all the layers are drawn and the model is complete, the model is then removed from the platform, and the support structures are removed from the part.

Parameters which affect performance and functionalities of the system are material column strength, material flexural modulus, material viscosity, positioning accuracy, road widths, deposition speed, volumetric flow rate, tip diameter, envelope temperature, and part geometry.

Materials

The materials that have been used for this additive manufacturing technique include polycarbonate (PC), Polyamide ((PA), acrylonitrile butadiene styrene (ABS), PC–ABS blends, medical-grade polycarbonate, wax, metals, and even ceramics.

Advantages

1) FDM process is able to fabricate prototypes with materials that are similar to that of the actual moulded product. This is especially useful in developing products that require quick prototypes for functional testing.

2) The FDM process build parts directly by extruding semi-liquid melt onto the model. Thus only those materials needed to build the part and its support are needed, and material wastages are kept to a minimum. There is also little need for cleaning up the model after it has been built.

3) This method is very convenient for users to get to their prototypes very quickly and there is very little or no post-processing necessary.

4) Build materials, supplied in spool form (or cartridge form in the case of the Dimension or Prodigy Plus), are easy to handle and can be changed readily when the materials.

Disadvantages

1) Parts built with the FDM process usually have restricted accuracy due to the shape of the material used, i.e., the filament form. Typically, the filament used has a diameter of 1.27 mm and this tends to set a limit on how accurately the part can be built.

2) The building process is slow, as the whole cross-sectional area needs to be filled with building materials. Building speed is restricted by the extrusion rate or the flow rate of the build material from the extrusion head.

3) As the FDM process extrudes the build material from its extrusion head and cools them rapidly on deposition, stresses induced by such rapid cooling invariable are introduced into the model. This leads to shrinkages and distortions of the models.

Applications

1) *Models for conceptualization and presentation:* Models can be marked, sanded, painted and drilled and thus can be finished to be almost like the actual product.

2) *Prototypes for design, analysis and functional testing:* The system can produce a fully functional prototype in ABS. The resulting ABS parts have 85% of the strength of the actual moulded part. Thus, actual testing can be carried out, especially with consumer products.

3) *Patterns and masters for tooling:* Models can be used as patterns for investment casting, sand casting and molding.

MATERIAL JETTING (METAL JET)

Material jetting creates objects in a similar method to a two-dimensional ink jet printer. Material is jetted onto a build platform using either a continuous or Drop on Demand (DOD) approach.

Material is jetted onto the build surface or platform, where it solidifies and the model is built layer by layer. Material is deposited from a nozzle which moves horizontally across the build platform. Machines vary in complexity and in their methods of controlling the deposition of material. The material layers are then cured or hardened using ultraviolet (UV) light.

As material must be deposited in drops, the number of materials available to use is limited. Polymers and waxes are suitable and commonly used materials, due to their viscous nature and ability to form drops.

Material Jetting Process

1. The print head is positioned above build platform.
2. Droplets of material are deposited from the print head onto surface where required, using either thermal or piezoelectric method.
3. Droplets of material solidify and make up the first layer.
4. Further layers are built up as before on top of the previous.
5. Layers are allowed to cool and harden or are cured by UV light. Post processing includes removal of support material.

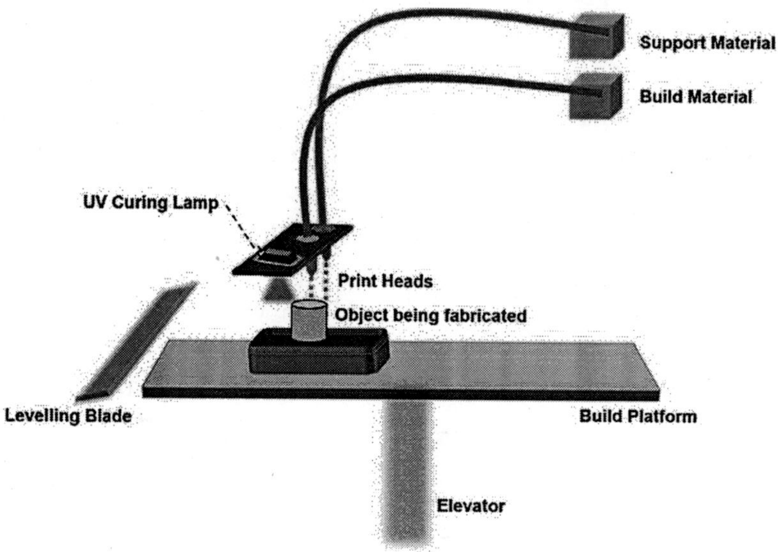

Figure 6.6 Material Jetting Process

Advantages:

1) The process benefits from a high accuracy of deposition of droplets and therefore low waste

2) The primary advantages of printing, both direct and binder printing, include low cost, high speed, scalability, ease of building parts in multiple materials, and the capability of printing colors.

3) Printing machines are much lower in cost than other AM machines, particularly the ones that use lasers.

4) In general, printing machines can be assembled from standard components (drives, stages, print heads), while other machines have many more machine-specific components.

5) High speed and scalability are related: by using print heads with hundreds or thousands of nozzles, it is possible to deposit a lot of material quickly and over a considerable area.

Disadvantages:

1) Support material is often required

2) The choice of materials to date is limited.

3) Only waxes and photopolymers are commercially available.

4) Part accuracy, particularly for large parts, is generally not as good as with some other processes, notably vat photo polymerization and material extrusion.

BINDER JETTING

The binder jetting process uses two materials; a powder-based material and a binder. The binder acts as an adhesive between powder layers. The binder is usually in liquid form and the build material in powder form. A print head moves horizontally along the x and y axes of the machine and deposits alternating layers of the build material and the binding material. After each layer, the object being printed is lowered on its build platform.

Due to the method of binding, the material characteristics are not always suitable for structural parts and despite the relative speed of printing, additional post processing (see below) can add significant time to the overall process.

As with other powder-based manufacturing methods, the object being printed is self-supported within the powder bed and is removed from the unbound powder once completed. The technology is often referred to as 3DP technology and is copyrighted under this name.

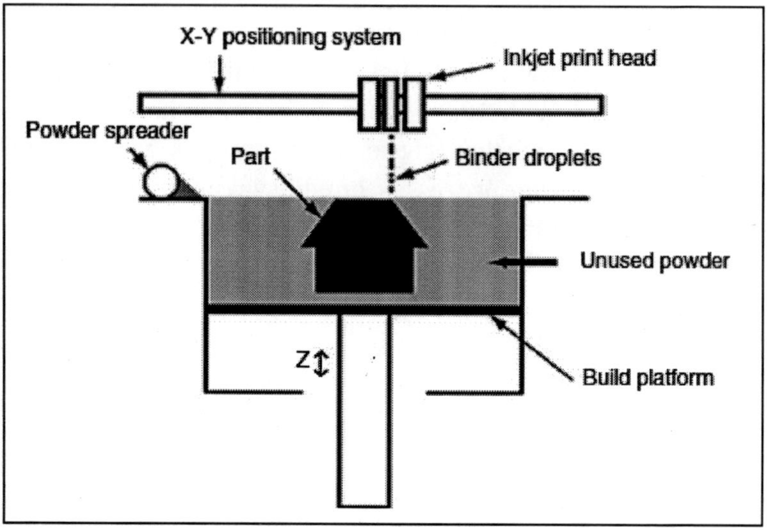

Figure 6.7 Binder Jetting Process

Binder Jetting Process

1. Powder material is spread over the build platform using a roller.

2. The print head deposits the binder adhesive on top of the powder where required.

3. The build platform is lowered by the model's layer thickness.

4. Another layer of powder is spread over the previous layer. The object is formed where the powder is bound to the liquid.

5. Unbound powder remains in position surrounding the object.

6. The process is repeated until the entire object has been made.

Advantages of Binder Jetting

1) Binder Jetting produces metal parts and full-color prototypes at a **fraction of the cost** compared to DMLS/SLM and Material Jetting respectively.

2) Binder Jetting can manufacture **very large parts** and **complex metal geometries**, as it is not limited by any thermal effects (e.g. warping).

3) The manufacturing capabilities of Binder Jetting are excellent for low to medium **batch production**.

Disadvantages of Binder Jetting

1) Metal Binder Jetting parts have **lower mechanical properties** than DMSL/SLM parts, due to their higher porosity.

2) Only **rough details** can be printed with Binder Jetting, as the parts are very brittle in their green state and may fracture during post processing.

3) Compared to other 3D printing process, Binder Jetting offers a limited material selection.

Chapter 7

AUTOMATION

7.1 INTRODUCTION

Today's highly increasing competitiveness over the industry demands high quality and most consistent products with a competitive price. To address this challenge number of industries considering various new product designs and integrated manufacturing techniques in parallel with the use of automated devices.

One of the remarkable and influential moves for getting the solutions of above-mentioned challenge is the industrial automation. ***Industrial automation*** facilitates to increase the product quality, reliability and production rate while reducing production and design cost by adopting new, innovative and integrated technologies and services

Automation takes a step further mechanization that uses a particular machinery mechanism aided human operators for performing a task. Mechanization is the manual operation of a task using powered machinery that depends on human decision making.

On the other hand, automation replaces the human involvement with the use of logical programming commands and powerful machineries.

In a brief, *industrial automation can be defined* as the use of set technologies and automatic control devices that results the automatic operation and control of industrial processes without significant human intervention and achieving superior performance than manual control. These automation devices include PLCs, PCs, PACs, etc. and technologies include various industrial communication systems.

7.2 TYPES OF AUTOMATION

Automation is technology concerned with the application of

Mechanical, electronic & computer-based systems to operate and control production in order to improves productions.

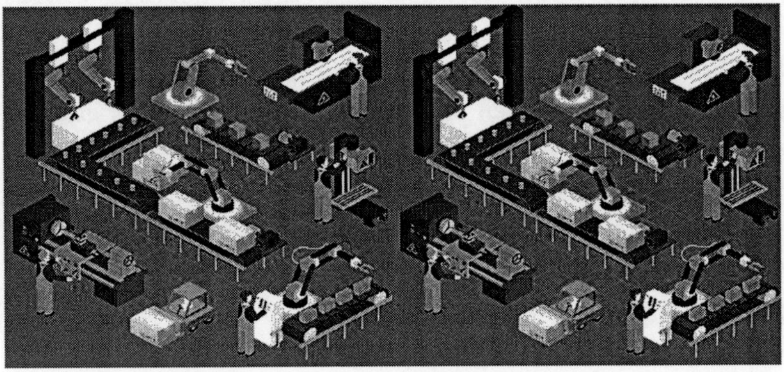

Figure 7.1 A typical Automated production system.

Types of Automation: Automated production systems are classified into three basic types:

1 Fixed automation

2 Programmable automation

3 Flexible automation

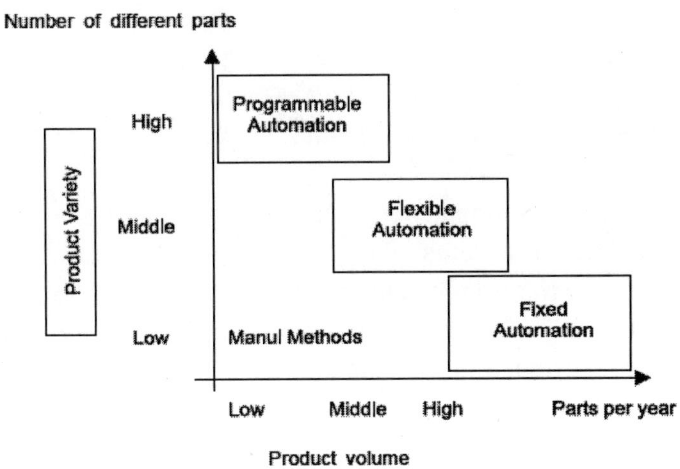

Figure 7.2 Types of Automation

1. Fixed automation

Fixed automation is a system in which the sequence of processing (or assembly) operations is fixed by the equipment configuration. The operations in the sequence are usually simple. It is the integration and coordination of many such operations into one piece of equipment that makes the system complex.

The typical features of fixed automation are:

- High initial investment for custom-engineered equipment
- High production rates
- Relatively inflexible in accommodating product changes

The economic justification for fixed automation is found in products with very high demand rates and volumes. The high initial cost of the equipment can be spread over a very large number of units, thus making the unit cost attractive compared to alternative methods of production.

2. Programmable automation

In programmable automation, the production equipment is designed with the capability to change the sequence of operations to accommodate different product configurations. The operation sequence is controlled by a program, which is a set of instructions coded so that the system can read and interpret them. New programs can be prepared and entered into the equipment lo produce new products. Some of the features that characterize programmable automation include:

- High investment in general-purpose equipment.
- Low production rates relative to fixed automation
- Flexibility to deal with changes in product configuration
- Most suitable for batch production

Automated production systems that are programmable are used in low and medium- volume production. The parts or products are typically made in batches. To produce each new batch of a different product, the system must be reprogrammed with the set of machine instructions that correspond to the new product. The physical setup of the machine must also be changed over: Tools must be loaded, fixtures must be attached to the machine table, and the required machine settings must be entered. This changeover procedure takes time. Consequently, the typical cycle for a given product includes a period during which the setup and reprogramming takes place, followed by a period in which the batch is produced.

3. Flexible automation

Flexible automation is an extension of programmable automation. The concept of flexible automation has developed only over the last 15 to 20 years, and the principles are still evolving. A flexible automated system is one that is capable of producing a variety of products (or parts) with virtually no time lost for changeovers from one product to the next. There is no production time lost while reprogramming the system and altering the physical setup (tooling, fixtures and machine settings). Consequently, the system can produce various combinations and schedules of products, instead of requiring that they be made in separate batches.

The features of flexible automation can be summarized as follows:

- High investment for a custom-engineered system
- Continuous production of variable mixtures of products
- Medium production rates
- Flexibility to deal with product design variations

The essential features that distinguish flexible automation from programmable automation are

(1) The capacity to change part programs with no lost production time, and

(2) The capability to change over the physical setup, again with no lost production time.

These features allow the automated production system to continue production without the downtime between batches that is characteristic of programmable automation. Changing the part programs is generally accomplished by preparing the programs off-line on a computer system and electronically transmitting the programs to the automated production system. Therefore, the time required to do the programming for the next job does not interrupt production on the current job. Advances in computer systems technology are largely responsible for this programming capability in flexible automation. Changing the physical setup between parts is accomplished by making the changeover off-line and then moving it into place simultaneously as the next part comes into position for processing. The use of pallet fixtures that hold the parts and transfer into position at the workplace is one way of implementing this approach. For these approaches to be successful, the variety of parts that can be made on a flexible automated production system is usually more limited than a system controlled by programmable automation.

The relative positions of the three types of automation for different production volumes and product varieties are depicted in Figure 9.3.

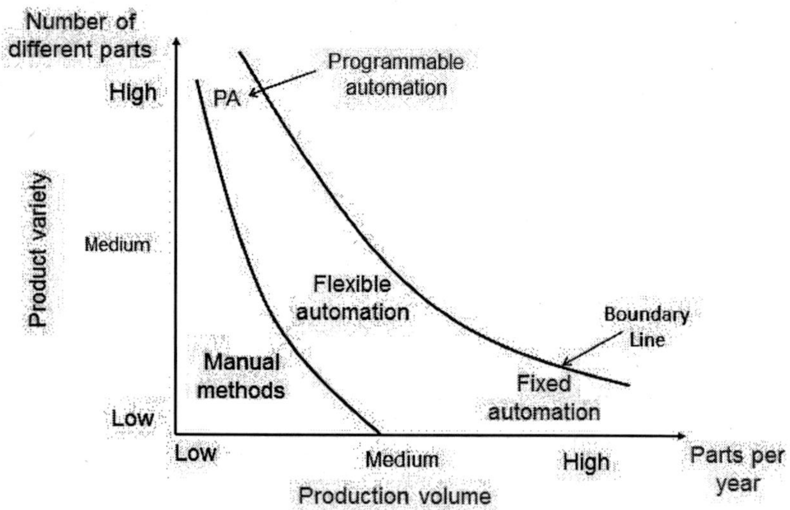

Figure 7.3 Automation Types against product variety v/s production volume.

7.3 REASONS FOR AUTOMATION

The important reasons for automation include the following:

1. ***Increased productivity***: Automation of manufacturing operations holds the promise of increasing the productivity of labor. This means greater output per hour of labor input. Higher production rates (output per hour) are achieved with automation than with the corresponding manual operations.

2. ***High cost of labor***: The trend in the industrialized societies of the world has been toward ever-increasing labor costs. As a result, higher investment in automated equipment has become economically justifiable to replace manual operations. The high cost of labor is forcing business leaders to substitute machines for human labor. Because machines can produce at higher rates of output, the use of automation results in a lower cost per unit of product.

3. ***Labor shortages:*** In many advanced nations there has been a general shortage of labor. Labor shortages also stimulate the development of automation as a substitute for labor.

4. ***Trend of labor toward the service sector:*** This trend has been especially prevalent in the advanced countries. First around 1986, the proportion of the work force employed in manufacturing stands at about 20%. In 1947, this percentage was 30%. By the year 2000, some estimates put the figure as low as 2%, certainly, automation of production jobs has caused some of this shift. The growth of government employment at the federal, state, and local levels has consumed a certain share of the labor market which might otherwise have gone into manufacturing. Also, there has been a tendency for people to view factory work as tedious, demeaning, and dirty. This view has caused them to seek employment in the service sector of the economy.

5. ***Safe:*** By automating the operation and transferring the operator from an active participation to a supervisory role, work is made safer. The safety and physical well-being of the worker has become a national objective with the enactment of the Occupational. Safety and Health Act of 1970 (OSHA). It has also provided an impetus for automation.

6. ***High cost of raw materials:*** The high cost of raw materials in

manufacturing results in the need for greater efficiency in using these materials. The reduction of scrap is one of the benefits of automation.

7. ***Improved product quality:*** Automated operations not only produce parts at faster rates than do their manual counterparts, but they produce parts with greater consistency and conformity to quality specifications.

8. ***Reduced manufacturing lead time:*** For reasons that we shall examine in sub sequent chapters, automation allows the manufacturer to reduce the time between customer order and product delivery. This gives the manufacturer a competitive advantage in promoting good customer service.

9. ***Reduction of in-process inventory:*** Holding large inventories of work-in-process represents a significant cost to the manufacturer because it ties up capital. In- process inventory is of no value. It serves none of the purposes of raw materials stock or finished product inventory. Accordingly, it is to the manufacturer's advantage to reduce work-in- progress to a minimum. Automation tends to accomplish this goal by reducing the time a work part spends in the factory.

10. ***High cost of not automating:*** A significant competitive advantage is gained by automating a manufacturing plant. The advantage cannot easily be demonstrated on a company's project authorization form. The benefits of automation often show up in intangible and unexpected ways, such as improved quality, higher sales, better labor relations, and better company image. Companies that do not automate are likely to find themselves at a competitive disadvantage with their customers, their employees, and the general public.

Review Questions

1. State the main objectives of a modern industry (at least five) and explain the role of automation in helping achieve them.
2. Differentiate fixed and programmable automation.
3. What are the principles of automation?
4. Explain ten Strategies of Automation?
5. What is industrial automation?

6. Explain the following types of Automation: (i) Programmable Automation (ii) Fixed Automation

7. Differentiate between fixed and flexible automation

8. Differentiate fixed and programmable automation.

9. What are the objectives of flow line automation?

10. Define `Fixed Automation' and `Flexible Automation'. Enumerate the differences between them.

Chapter 8

Civil Engineering and its Application

8.1 Introduction

The basic needs of the human being are water, food, clothing and shelter. With rapid industrialization and urbanization, fresh and unpolluted air has become the added necessity of the urbanites. To fulfil these needs, different works are executed. Engineering is comprised of the execution of such works. The branch of knowledge which deals with various industrial processes is defined as the technology. The technology is a systematic knowledge of industrial processes.

Engineering can be defined as the role of directing the resources of nature to the use and convenience of man. Technology is an application of basic scientific knowledge to advancement.

Branches of Engineering

The major branches of engineering are:

1. Civil Engineering
2. Mechanical Engineering
3. Electrical Engineering 4. Electronics
5. Telecommunication Engineering
6. Computer Engineering
7. Information Technology
8. Instrumentation
9. Metallurgy
10. Chemical Engineering
11. Mining

Civil Engineering

It is the first basic branch of engineering. Its relation to the fulfilment of human needs is direct, whereas the other engineering branches are complementary to the function of civil engineering.

Civil engineering is called the general engineering because civil engineering uses the principles and information provided by other branches of engineering while erecting civil engineering structures.

8.2 Civil Engineering for Civilization

Civilization is not built by bricks and mortar but by the men and women. Civilization is the average intellectual level of this population based on their educational standards and outcomes there of to analyse a given problem on happening of a thing/event in a correct perspective.

In the olden days, the word Military Engineer originated first as he undertook the engineering

tasks with reference to military's role in the war. The engineer undertaking works other than military engineering was referred to as a Civil Engineer. By civil engineering, we mean the engineering know-how and the tasks undertaken for the purpose of the civil population. Engineer and a civil engineer lay in the TIME resource. Different branches of civil engineering are explained below.

Surveying

The study of surveying enables the engineer to determine relative positions of points on the surface of earth. Before the construction of any structure, surveying of the land, on which the construction is to come up, is always done. From the field observations in surveying taken in the horizontal plane, a plan is prepared of the existing features of ground and relative positions of points in the vertical plane are shown by sections, obtained by taking measurements in the vertical plane, termed levelling.

Classification of surveying

Surveying is classified into two broad categories:

1. Plane surveying
2. Geodetic surveying.

Plane surveying: In this type of surveying, the effect of curvature of the surface of earth is not taken into account. The surface of earth is taken to be a plane surface. The extent of area up to which this assumption of plane surface holds good is 260 sq.km. Therefore, plane surveying is adopted for small areas. The degree of accuracy of plane survey is relatively low.

Geodetic surveying: In geodetic surveying the curvature of the earth's surface is taken into account. Geodetic surveys are carried out for large distances and areas with a high degree of precision. In India, Survey of India, a Government department, undertakes the work of geodetic surveying.

General classification of surveying

Surveys are generally classified on the basis of purpose of the survey and the instruments used for surveying.

Classification based on the purpose of surveying

Topographical surveys: These surveys are carried out to locate the natural and artificial features on the surface of earth, such as rivers, hills, lakes, roads, railways, towns, villages, etc.

Cadastral surveys: These surveys are required for determining the boundaries of fields, houses and other properties to a larger scale.

City surveys: These surveys are carried out for laying out the boundaries of plots for the location of water supply and sewer lines.

Mine surveys: In this survey the underground mineral wealth such as iron ores, coal, gold, etc.are explored.

Classification based on the instrument used for surveying

1. Chain survey

2. Compass survey

3. Chain and compass survey

4. Theodolite survey

5. Tachometric survey

6. Plane table survey

7. Electronic distance meter survey.

Purpose of surveying

1. Surveying is necessary for preparing the plans which are the projections of any object on the horizontal plane. The differences in height are denoted by contour lines.

2. Surveying is necessary for planning the design of any engineering work. The accuracy

of surveying influences, the success and governs the economy of the project.

8.3 Building materials

The materials required for the construction of structures are called either civil engineering materials or building materials. It is very essential for an engineer, builder, architect and contractor to be thoroughly conversant with these building materials. The knowledge of different types of materials, their properties and uses for different purposes is very essential for the builder in achieving overall economy.

Building materials account for about 70% of the total cost of construction. Thus, it is important that the building materials are easily and cheaply available.

Some of the building materials are:

1. Bricks: A brick is a rectangular block of regular shape obtained by moulding the mixture of clay and sand, which is then generally burnt at a high temperature.

2. Rubble or stone: A building stone or rubble is a natural material quarried from geological rock formations of igneous, sedimentary or metamorphic type. When quarried, stones are irregular in shape and have rough surface. Irregular stones are brought to the required size and shape and the process is called the dressing of stone. These dressed stones or rubbles are used in stone masonry.

3. Aggregates: Aggregates are chemically inert materials, such as crushed stone, gravel,sand, broken bricks, blast furnace slag, etc., obtained naturally or by crushing.

4. Cement: Cement is a binding material used in preparing cement mortar or concrete.

5. Alternative building materials: Traditional materials have a limited use for heavy constructions and therefore there is a need to develop new materials. Hence to reduce the cost of construction and to increase the strength, alternative building materials are used, for example, mud blocks, concrete blocks (solid or hollow), ferro cement, etc.

6. Composite materials: When two or more materials are combined to act as a single material, it is called the composite material, for example, wood laminates, plastic laminates, asbestos cement sheet, reinforced glass, etc.

7. Smart materials: These are materials which exhibit a change in their non-mechanical properties like thermal, optical, electrical or magnetic because of changes in the mechanical stimulus.

Construction Technology—Built Environment

A suitable environment is created by constructing a building. The building technology covers the planning of different units of a building to provide a suitable environment for the activities designed for the building. Codes of building bye-laws ensure good and sound construction through regulating the materials and construction methods. Climatic influence on the built environment is vital for minimum energy consumption. The building technology also covers the maintenance and repairs of the buildings and their safe demolition when they become too old to be used further.

The buildings are classified according to functions such as:

1. Residential buildings

2. Public buildings

3. Commercial buildings

4. Industrial buildings

5. Recreation buildings

6. Educational buildings

7. Hospital buildings

8. Storage, i.e. warehouses, etc.

9. Special purpose buildings, non-conventional buildings.

Building technology deals with the analysis and design of substructures as well as

super structures of buildings. It includes the study of different construction materials in respect of their properties and construction techniques. Some of the building materials are metals, timber, concrete, bituminous materials, bricks, soil, polymers and plastics, etc.

Construction technology and its management The scope of 'construction' is more comprehensive here than that is in building technology. Based on sound principles of soil mechanics, foundations to non-conventional structures are covered under construction technology. It is comprised of different techniques of construction for different materials under different site conditions. The study of construction machinery comes under its

purview. The management or organization of men (labour), materials, methods in relation to site, money and time is the backbone of construction management. It involves almost every branch of engineering, commerce and economics, for its ultimate aim is to 'achieve the desired construction in the most economical way'. A clear knowledge of the following points is necessary for reliable construction and its management.

The ten Ms of construction management are:

(a) Money, Materials, Machines, Manpower, Methodologies,

(b) Maintenance, Modernization, Monitoring, Motivations,

(c) Managements of all types.

8.4 Geotechnical Engineering

This branch of civil engineering is also called soil mechanics. It is a discipline of civil engineering in which the study of soil, its behaviour on the application of load and its use as an engineering material in the construction of earth dams, is done. The properties and strength characteristics of different types of soil are studied in this subject. The knowledge of this subject is useful in the design of earth dams, different pile foundations, buildings, foundations.

A building's foundation transmits loads from buildings and other structures to the earth. Geotechnical engineers design foundations based on the load characteristics of the structure and the properties of the soils and/or bedrock at the site. In general, geotechnical engineers:

1. Estimate the magnitude and location of the loads to be supported.

2. Develop an investigation plan to explore the subsurface.

3. Determine necessary soil parameters through field and lab testing

4. Design the foundation in the safest and most economical manner.

Offshore

Offshore (or marine) geotechnical engineering is concerned with foundation design for human-made structures in the sea, away from the coastline (in opposition to onshore or near shore). Oil platforms, artificial islands and submarine pipelines are examples of such structures. There are a number of significant differences between onshore and offshore geotechnical engineering. Notably, ground improvement (on the seabed) and site investigation are more expensive, the offshore structures are exposed to a wider range of geo hazards, and the environmental and financial consequences are higher in case of failure. Offshore structures are exposed to various environmental loads, notably wind, waves and currents. These phenomena may affect the integrity or the serviceability of the structure and its foundation during its operational lifespan – they need to be taken into account in offshore design.

In subsea geotechnical engineering, seabed materials are considered a two-phase material composed of 1) rock or mineral particles and 2) water.

Structures may be fixed in place in the seabed as is the case for piers, jettys and fixed-bottom wind turbines or maybe a floating structure that remains roughly fixed relative to its geotechnical anchor point.

Structural engineering

Structural engineering is the field of engineering that deals with the structural integrity and strength of a building or structure. Structural engineering is a specialty of civil engineering that ensures the structures are safe, stable and don't collapse under applied loads. It is mainly focused on analysis and design of the structure.

Analysis of the structure

- Careful analysis of the wind speed that can carry structural loads and the overall capacity and utility of the building also provides information.
- Analysis of the structure according to the principles of structural engineering will make sure that the structure depends on all the necessary design codes.

Design of the structure

- Structures have to be designed so that they can withstand their own weight as well as the loads and pressures that will be placed upon them.
- Structural engineers take steps crucial information about the foundations, roof types, load types, beams, columns, material quality, retaining walls etc.

Responsibilities of a structural engineer

A structural engineer also plays a major role as a team among other professionals like surveyor, quantity surveyor, and architects engineers.

The following tasks must be performed by a structural engineer:

- Design models of structures using software.
- Assessing the reaction of structures to pressures and stress.
- Finalizing the appropriate concrete materials that would be suitable for the structure.
- Assessing budget of the project.
- Liaising to ensure that newly erected buildings are structurally sound with construction contractors.
- Using computers and computer-aided design technology for simulation purposes.
- Knowledge of CAD software application like ETABS, SAFE, STAAD Pro, PROKON, REVIT, ROBOT etc.

Role of Structural Engineering in building design & sustainable development

- Urbanization, buildings, and civil infrastructure are most important of the world for any economy. Require lots of natural resources and energy required for their construction, operation, maintenance and deterioration.

- Structural engineers play an important role in the efficient design, construction and execution of a building. At the beginning of construction projects in collaboration with structural engineers, architects, owners and construction managers, architects decide on the effectiveness, feasibility and cost-effectiveness of concepts. As the project progressed to the design stage, structural engineers produced the first set of executive design documents.
- Structural engineers have traditionally not identified themselves with environmental sustainability. Growing concerns for the natural environment, the rise of smart cities, and the growing importance of sustainability and energy efficiency are driving the demand for structural engineers.
- Building construction consumes natural resources, potable water and energy. Secondly, the construction process also contributes to 30% of the global greenhouse gas (GHG) emissions. Thus, a sustainable environment is not achievable without contribution of the building & construction sector, especially the structural engineering domain.
- The renewable energy department is a great leading factor. The concept of sustainability is gaining popularity these days and it is clearly increasing the applicability in the energy sector.

8.5 Transportation Engineering

This subject deals with the transport of men and materials through different communication routes such as land, water and air. The railways and roads are the important modes of communication by land. The water transport is feasible only where the rivers, canals are navigable or where the seacoast is available. Transportation by air routes is also increasing day by day. The transportation of men and goods for short distances can be speedily done by road transport, whereas railway transport is considered safe and convenient for long distances. The knowledge of surveying and levelling is very useful before deciding the alignments of roads or railway. The preparation of contour plan of a hilly region obtained from a levelling operation is useful in deciding the alignment of hill roads.

The knowledge of surveying and levelling enables the engineer to decide the alignment of tunnels, which become necessary when the road or railway transportation is to be done through the hilly regions. Transportation engineering is further classified into:

1. Highway engineering

- Handle the planning, design, construction, and operation of highways, roads, and other vehicular facilities as well as their related bicycle and pedestrian realms
- Estimate the transportation needs of the public and then secure the funding for projects
- Analyze locations of high traffic volumes and high collisions for safety and capacity
- Use engineering principles to improve the transportation system
- Utilize the three design controls, which are the drivers, the vehicles, and the roadways themselves

2. Railroad engineering

Railway engineers handle the design, construction, and operation of railroads and mass transit systems that use a fixed guideway (such as light rail or monorails). Typical tasks include determining horizontal and vertical alignment design, station location and design, and construction cost estimating. Railroad engineers can also move into the specialized field of train dispatching which focuses on train movement control.

Railway engineers also work to build a cleaner and safer transportation network by reinvesting and revitalizing the rail system to meet future demands.

3. Port and harbour engineering

Port and harbour engineers handle the design, construction, and operation of ports, harbours, canals, and other maritime facilities.

4. Airport engineering

Airport engineers design and construct airports. Airport engineers must account for the impacts and demands of aircraft in their design of airport facilities. These engineers must use the analysis of predominant wind direction to determine runway orientation, determine the size of runway border and safety areas, different wing tip to wing tip clearances for all gates and must designate the clear zones in the entire port.

8.6 Hydraulics and Water Resources Engineering

Water is such a commodity that it is vital for the existence of mankind. Human beings, animals and plants require water for their survival. Surface water is easy and economical to harness, however, its availability cannot be relied upon continuously since it varies with the season.

Hydraulics is a branch of science in which the study of fluids, i.e. liquids and gases, at rest and in motion is done. Usually, the liquid is water and the subject is titled *hydraulics*. When the water is as rest, the forces exerted by the water on immersed areas are found out by the laws of mechanics. Thus, the knowledge of these forces is useful in the design of the gates used to control the flood water in case of dams. When water is drawn off from a reservoir and conveyed through closed conduits or open channels, the knowledge of the behaviour of liquids in motion is useful here. Thus in the design of water supply distribution systems, the study of fluid mechanics helps to solve the problems encountered in the design. Certain machines which work on the water are used for the generation of electricity and are termed hydraulic machines. Knowledge of fluids mechanics is useful in designing these machines so that they give the best possible output.

Water Resources Engineering can be defined as the science which deals with the subject of tapping water either from the surface or subsurface sources. A water resource is such a vast subject that it includes in itself hydrology, irrigation, hydraulics and water supply. Tremendous volume of water is stored in the earth's crust. According to one estimate the total volume stored under the surface of earth maybe about 80 million km cube, half of which may be at depths less than 600 m. The use of surface water for irrigation is likely to create problems like water logging in certain areas. Groundwater is obtainable all the year round and its use along with the surface water keeps the subsoil water level within reasonable limits. The judicious use of water for purposes of irrigation has gained such importance in the recent times that water management has become a science in itself. Engineers have learnt to tame the water resources by construction of dams, construction of bore wells and construction of hydroelectric plants for the benefit of mankind.

8.7 Environmental Engineering

This is an important branch of civil engineering which covers both water supply and sanitary engineering. The importance of clean environment was felt with the rapid growth in population, and the growth in urbanization and industries. Environment is polluted through the mediums of air, water or such other agents. The science of civil engineering deals with the subject of tapping water from different sources, testing its quality, purification processes and distribution of water to the consumers. Similarly, the environmental engineering encompasses the subject of treatment of wastes which originate from different sources and deals with the removal of harmful substances in these wastes by different processes. The impact of wastes originating from industries is felt by living organisms if such wastes contain toxic substances. The Central Government as well as state governments have enacted laws for the protection of environment needed for the safe living of human beings.

Applications

1. Water supply and treatment

Environmental engineers evaluate the water balance within

a watershed and determine the available water supply, the water needed for various needs in that watershed, the seasonal cycles of water movement through the watershed and they develop systems to store, treat, and convey water for various uses.

Water is treated to achieve water quality objectives for the end uses. In the case of a potable water supply, water is treated to minimize the risk of infectious disease transmission, the risk of non-infectious illness, and to create a palatable water flavor. Water distribution systems are designed and built to provide adequate water pressure and flow rates to meet various end-user needs such as domestic use, fire suppression, and irrigation.

2. Wastewater treatment

There are numerous wastewater treatment technologies. A wastewater treatment train can consist of a primary clarifier system to remove solid and floating materials, a secondary treatment system consisting of an aeration basin followed by flocculation and sedimentation or an activated sludge system and a secondary clarifier, a tertiary biological nitrogen removal system, and a final disinfection process. The aeration basin/activated sludge system removes organic material by growing bacteria (activated sludge). The secondary clarifier removes the activated sludge from the water. The tertiary system, although not always included due to costs, is becoming more prevalent to remove nitrogen and phosphorus and to disinfect the water before discharge to a surface water stream or ocean outfall.

3. Air pollution management

Scientists have developed air pollution dispersion models to evaluate the concentration of a pollutant at a receptor or the impact on overall air quality from vehicle exhausts and industrial flue gas stack emissions. To some extent, this field overlaps the desire to decrease carbon dioxide and other greenhouse gas emissions from combustion processes.

4. Environmental impact assessment and mitigation

Environmental engineers apply scientific and engineering

principles to evaluate if there are likely to be any adverse impacts to water quality, air quality, habitat quality, flora and fauna, agricultural capacity, traffic, ecology, and noise. If impacts are expected, they then develop mitigation measures to limit or prevent such impacts. An example of a mitigation measure would be the creation of wetlands in a nearby location to mitigate the filling in of wetlands necessary for a road development if it is not possible to reroute the road.

In the United States, the practice of environmental assessment was formally initiated on January 1, 1970, the effective date of the National Environmental Policy Act (NEPA). Since that time, more than 100 developing and developed nations either have planned specific analogous laws or have adopted procedure used elsewhere. NEPA is applicable to all federal agencies in the United States.

8.8 Architecture and Town Planning

The fundamental process of planning, designing, constructing structures or buildings is known as architecture. It is sometimes referred to as an art because a piece of construction eventually becomes a part of the cultural heritage. However, as a field of study, architecture and town planning is both art and science as it requires both artistic skills and knowledge of science. Urban planning, also termed as town planning, is more of a technical process of design and development aimed at optimal use of land and building of infrastructure in public areas. The most important concern out here is public welfare along with the protection of the environment. The planning mostly helps to build rural and suburban areas, settling up communities by facilitating the supply of water and other resources. Regional planning is nothing but urban planning but on a broader scale. It covers aspects like the optimal use of land, protection of farmland, creation of industrial space, transportation hubs and cities, etc. The basic principle of architecture and town planning at the macro level is to resist any kind of development in the floodplains and earthquake vaults and utilize them in some other ways. Allocation of land for different purposes, laying down regional level policies and even-zoning of laws and policies for the best use of the land- all come under the purview of this course.

8.9 AI in Civil Engineering.

Artificial Intelligence in Civil Engineering works on the goal of imitating and executing functions of human brain, logically and intelligently. According to Robert J. Schalkoff in his book 'Artificial Intelligence Engine' (January 1990), artificial intelligence is "a field of study that seeks to explain and emulate intelligent behaviour in terms of computational processes". The concept of artificial intelligence is extensively used in the field of construction.

Artificial intelligence work based on different methods: These can be fuzzy systems, Neural networks, Knowledge based systems or genetic algorithms.

4.2.1 Applications Artificial Intelligence in Civil Engineering & Construction

Each and every civil engineering project is associated with risks and uncertainties. This can include the risks regarding task force allocation, mile stone achievement, project costing and overall construction management. Machine learning which a branch of Artificial Intelligence (AI) is is widely used in the domain of civil engineering.

The following are few applications of artificial intelligence in civil engineering:

1. For estimating the percentage of soil moisture content and further classifications.

2. In the structural engineering field machine learning can be applied to detect damages using sensory or image data, identifying it's location and extent.

3. Improving productivity by reducing idle time.

4. For predicting maximum dry density and optimum moisture content in concrete.

5. Using image recognition for proper site monitoring, including aspects of safety and dangerous working conditions.

6. Identifying gaps and requirement of materials to cover the tasks without delay.
7. For travel time prediction and sign AI optimization in transportation engineering.
8. Efficient planning, designing and managing of infrastructure using Building Information Modelling (BIM).
9. Utilizing Artificial Neural Network for predicting properties of concrete mix designs.
10. To monitor activity in the construction site and predicting changes in the costing based on raw material market rates.
11. To analyse settlement of foundation and slope stability.
12. For monitor real time structural health of the building, giving warnings on when and where repair is required.
13. Helping in tidal forecasting to aid construction in marine environment.
14. Reducing errors in the project by automatic analysis of data.
15. To develop site layouts and predict risks as part of project management.
16. Finding a solution for damage related to pre-stressed concrete pile driving in foundation engineering.
17. To solve complicated problems in different stages of the project.

8.10 Geographic Information Systems (GIS) and Remote Sensing:

The data that GIS uses to create maps can originate from a virtually limitless number of sources. From worksheets to satellites, IoT to information databases, GIS can turn any source of relevant data into an interactive map. One of its most remote sources of data for GIS is remote sensing data. Remote sensing is the process of gathering data from afar. Drones, aircraft, and satellites can carry sensors that can make maps using several sectors of the electromagnetic spectrum. LiDAR is one such example. Others are passive and use reflected sunlight, since some project various wavelengths of electromagnetic energy to the target and analyze the reflected signals. Faraway sensing also makes it possible to examine remote locations that would otherwise be unreachable.

Applications in Civil Engineering

1. Structural

GIS allows civil engineers to include a variety of material data and area historical data into their layouts. As an outcome, one of the most extensively utilized GIS applications is structural analysis. By merging 3D GIS maps with normal design techniques, designs can gain from previous mistakes. GIS mapping has a significant advantage over tabular data.

2. Environmental

GIS can assist organizations in need of environmental information on land, water supplies, and other natural elements. GIS software may create as many overlay map layers as needed to provide engineers with enough data to conduct impact assessments. In analysis maps, statistical data, future growth plans, industry issues, and other data can all be merged. Engineers can make the most environmentally friendly decisions by considering all of the aspects that can harm the environment at the same time. Patterns in natural events such as soil erosion can be discovered using the data. It can also help with natural disaster forecasts.

3. Transportation

Transportation is another clear sector where GIS can be useful due to its capacity to aid with road planning and logistics. There were also advancements in GPS technology that incorporates GIS technologies. Using GIS techniques, traffic flow trends can be displayed alongside demographic transition on the same map. New map layers, such as those showing the optimum courses for future bridges, can be updated at any time. GIS applications are similarly well-built. GIS software can handle highly dynamic traffic data as well as changing quickly flood levels.

4. Wastewater/Stormwater

GIS techniques can be used to combine data from hydraulic and hydrologic modeling to give a full examination of water utility systems. To forecast water demand, GIS incorporates data from customer information systems, water flow at different nodes, and historical data. Another application is to use terrain analysis and floods data to plan drainage improvements.

5. Site Analysis

GIS quickly analyses and combines a wide range of pictures and data. Because clients may easily relate to a simple map, the data is extremely accurate and spatially presented. The base map can be enhanced with serial pictures, environmental protection regions, city and zoning designations, soil and topographic maps, and aerial images. Overlays of important data on economic activity, transportation flow, and population growth combine to immediately present a clear picture of the site's constraints and opportunities

6. Surveying

Using GIS technologies, surveyors may quickly examine environmentally sensitive areas, forestry, government control, road networks, previously established boundaries, zoning, permit status, and other important data. Because exact measurements are needed to construct accurate mapping systems, GIS proves to be an indispensable tool in surveying. These enhanced capabilities eliminate duplication of work and promote collaboration with other planning and government agencies. The information is preserved and easily accessible for future projects that utilize the area.

7. CAD Integration

Web pages are utilized to access files on national GIS servers, and level CAD documents are adapted and maintained from a central place, reducing duplicate datasets and providing a platform for all spatial data supply and functionality.

8.12 Innovative materials and practices in construction technology

Pre-Engineered Buildings (PEB'S)

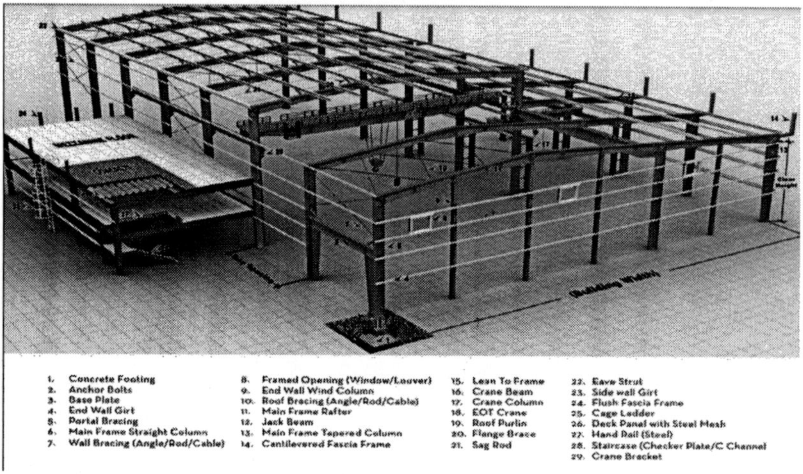

Figure 8.1 PEB Skeleton Structure

Technological improvement over the year has contributed immensely to the enhancement of quality of life through various new products and services. One such revolution was the pre-engineered buildings. Through its origin can be traced back to 1960's, its potential has been felt only during the recent years. This was mainly due to the development in technology, which helped in computerizing the design.

It is very advantageous over the conventional buildings and is really helpful in the low-rise building design. Pre-engineered buildings are generally low-rise buildings however the maximum eave height can go up to 25 to 30 metres. Low rise buildings are ideal for offices, houses, showrooms, shop fronts etc. The application of pre-engineered buildings concept to low rise buildings is very economical and speedy. Buildings can be constructed in less than half the normal time especially when complemented with the other engineered sub systems. The roof of low-rise buildings may be flat or sloped. Intermediate floors of low-rise buildings are made of mezzanine systems. Single storied houses for living take minimum time for construction and can be built in any type of geographical location like extreme cold hilly areas, high rain prone areas, plain land obviously and extreme hot climatic zones as well.

Applications of Pre-Engineered Buildings (PEB'S)

1. Warehouses
2. Factories
3. Workshops
4. Offices
5. Gas stations
6. Vehicle parking sheds
7. Showrooms
8. Aircraft hangars
9. Metro stations
10. Schools
11. Indoor stadium roofs
12. Outdoor stadium canopies
13. Bridges
14. Railway platform shelters

Advantages of Pre-Engineered Buildings (PEB'S)

Reduced Construction Time: Buildings are typically delivered in just a few weeks after approval of drawings. The use of PEB will reduce the total construction time of the project by at least 50%. This also allows faster occupancy and earlier realization of revenue.

Lower cost: Due to the systematic approach, there is a significant saving in design, manufacturing and on-site erection cost. The secondary members and cladding nest together reducing transportation cost.

Flexibility of expansion: Buildings can be easily expanded in length by adding additional bays. Also, expansion in width and height is possible by pre designing for future expansion.

Quality control: As buildings are manufactured completely in the factory under controlled conditions the quality is assured.

Single source responsibility: As the complete building package is supplied by a single vendor, compatibility of all the building components and accessories is assured. This is one of the major benefits of the pre - engineered building systems.

Hybrid Construction

The term 'hybrid concrete construction' (HCC) is used to describe the combination of concrete construction methods, e.g. in-situ concrete concrete construction' and precast concrete, with one another or with other materials, e.g. steel.

Hybrid concrete construction is the combination of precast and insitu concrete. Different materials are used according to their strengths and weaknesses to provide simple, buildable and competitive high-quality structures that offer consistent performance. Hybrid concrete construction can incorporate all the benefits of precasting, (e.g. form, finish, colour, speed, accuracy, prestressing, high-quality, assured covers and dense and properly cured coversheet) with all the benefits of in-situ construction (e.g. economy, flexibility, malleability, continuity and robustness).

Figure 8.2 Hybride Structure

Benefits of using Hybrid Concrete Construction

1. Time and Cost of Construction

Construction of hybrid concrete structures provides good quality and greater speed during construction making it more economical. Thus, cost of hybrid concrete construction is less compared to normal construction.

2. **Buildability of Hybrid Concrete**

Hybrid concrete structural elements are manufactured by skilled workforce. Thus, quality of structural members is good. The use of this concrete also helps to reduce time of construction as design and construction decisions are taken before the start of construction during design stage itself.

3. **Safety during Construction**

As most of the hybrid concrete structural members are constructed at factory, it offers more safety to workforce during construction process. The plan for safety procedures for installation of these members can be made based on site conditions and in advance, reducing the chance of accidents at construction site.

3D printing in construction

Figure 8.3 3-D Printed Structure

Introduction

3D printing (sometimes referred to as Additive

Manufacturing (AM)) is the computer-controlled sequential layering of materials to create three-dimensional shapes. It is particularly useful for prototyping and for the manufacture of geometrically complex components.

Construction_industry

In the construction industry, 3D printing can be used to create construction components or to 'print' entire buildings. Construction is well-suited to 3D printing as much of the information necessary to create an item will exist as a result of the design process, and the industry is already experienced in computer aided manufacturing.

Construction 3D printing may allow, faster and more accurate construction of complex items as well as lowering labour costs and producing less waste. It might also enable construction to be undertaken in harsh or dangerous environments not suitable for a human workforce such as in space.

Recently, the 3D-printed building industry has seen some significant milestones. The largest 3D-printed building in the world was erected in Dubai at the beginning of 2020. An administrative office building, it spans more than 6,900 square feet.

Types of Materials

While 3D-printed buildings can be made from a number of different components, the most common method involves **a mix of concrete, fiber, sand, and geopolymers.**

These different raw materials are thoroughly mixed in a large "hopper," at which point they can be fed into the extrusion apparatus and layered into the correct shapes and patterns. There have also been instances of homes 3D printed from **fully biodegradable materials, including mud, soil, straw, and rice husks.**

Challenges to 3D-Printed Buildings

While 3D printing provides civil engineers with a number of opportunities, it also comes with challenges. Specifically, engineers may face hurdles related to structural inhabitability, material integrity, and equipment limitations.

1. **Structural Inhabitability**

One longstanding concern with 3D-printed buildings is that 3D printing technology is not, in and of itself, enough to make buildings habitable. While 3D printers can make walls and frames, they have historically come up short in terms of quality-of-life essentials such as infrastructure for running water and electricity work.

In recent years, 3D printing technology has made some strides, with new methods being developed for printing some basic electrical and plumbing components. Additionally, engineers can design homes in such a way that the basic 3D-printed structures are easily retrofitted with traditional pipes, wiring, and ducts.

2. **Material Integrity**

An even more serious drawback is that 3D-printed structures, when tested, usually turn out to be less strong and sturdy than more conventional buildings. This is because the materials used in 3D-printed buildings tend to break down and lose some of their structural integrity over time.

Engineers have long been working on new material methods that can withstand rigorous testing and comply with all pertinent building safety codes. There have already been many successes in this regard, including refining existing concrete and polymer mixes and incorporating biodegradable materials.

3. **Equipment Limitations**

It should also be noted that, while 3D printers have come a long way, the equipment itself still has limitations. While one of the great promises of 3D printers is that they can do the work of many machines at once, the reality is that a lot of 3D printers are restricted in their functionality. This means they can make large structures, but not necessarily complex or detailed ones.

Cement replacing materials

Sustainability is an important issue all over the world. Carbon dioxide emission has been a serious problem in the world due to the greenhouse effect. Today many countries agreed to reduce the emission of CO_2. Many phases of cement and concrete technology can affect sustainability. Cement and concrete industry is responsible for the production of 7 % carbon dioxide of the total world CO_2 emission. The use of supplementary cementing materials (SCM), design of concrete mixtures with optimum content of cement, and enhancement of concrete durability are the main issues toward sustainability in concrete industry

1) **Fly ash**: Fly ash is a by-product of the combustion of pulverized coal in thermal power plants. The dust-collection system removes the fly ash, as a fine particulate residue, from the combustion gases before they are discharged into the atmosphere. Fly ash particles are typically spherical, ranging in diameter from\1 lm up to 150 lm. The type of dust collection equipment used largely determines the range of particle sizes in any given fly ash. The fly ash from boilers at some older plants using mechanical collectors alone is coarser than from plants using electrostatic precipitators

2) **Granulated Blast Furnace Slag:** Metallurgical industry produces slag as by-products. Iron blast furnace slag is the major non-metallic product consisting of silicates and aluminosilicates of calcium. They are formed either in glassy texture used as a cementitious material or in crystalline forms used as aggregates. Other slags such as copper slag have pozzolanic properties and react with lime. Steel slags are usually produced in crystalline form and are used as base materials for road construction or as aggregates in special concrete productions. The other utilizations of slags are in the production of slag wool for thermal isolation in the building industry and as lightweight aggregates for lightweight concretes.

3) **Silica fume**: Silica fume or microsilica is very fine non-crystalline silica produced in electric arc furnaces as a by-product of the production of elemental silicon or alloys containing silicon. Micro-silica was first tested in concrete in Norway in the early 1950s. Higher strength was obtained for concretes containing silica fume. Performance of silica fume concretes in sulphate environment was

also better than normal Portland cement concretes.

4) Metakaolin :Metakaolin (MK), commercially available since the mid-1990s, is one of the recently developed supplementary cementing materials.Metakaolin differs from other supplementary cementitious materials (SCMs), like fly ash, silica fume, and slag, in that it is not a by-product of an industrial process; it is manufactured for a specific purpose under carefully controlled conditions. This allows manufacturing process of metakaolin to be optimized, ensuring the production of a consistent pozzolanic material. Metakaolin is produced by heating kaolin, one of the most abundant natural clay minerals, to temperatures of 650–900 °C. The *Meta* prefix in the term is used to denote change. The scientific use of the prefix is used for a combining form denoting the least hydrated of a series. In the case of metakaolin, the change that is taking place is dehydroxylization, brought on by the application of heat over a defined period of time. This heat treatment, or calcinations, serves to break down the structure of kaolin. Bound hydroxyl ions are removed and resulting disorder among alumina and silica layers yields a highly reactive, amorphous material with pozzolanic and latent hydraulic reactivity, suitable for use in cementing applications

5) Rice Husk ash: Rice-husk (RH) is an agricultural by-product material. It constitutes about 20 % of the weight of rice. It contains about 50 % cellulose, 25–30 % lignin, and 15–20 % of silica. When rice-husk is burnt rice-husk ash (RHA) is generated. On burning, cellulose and lignin are removed leaving behind silica ash. The controlled temperature and environment of burning yields better quality of rice-husk ash as its particle size and specific surface area are dependent on burning condition. For every 1000 kg of paddy milled, about 200 kg (20 %) of husk is produced, and when this husk is burnt in the boilers, about 50 kg (25 %) of RHA is generated. Completely burnt rice-husk is grey to white in color, while partially burnt rice-husk ash is blackish

Manufactured Sand (M-Sand)

Fine and coarse aggregate constitute about 75 % of total volume of concrete. The most commonly used fine aggregate is natural river sand. Nowadays the demand for river sand is increasing due to its lesser availability. Sand quarrying has resulted in scarcity and poses environmental problems such as loosing water retaining sand strata, deepening of the river courses and causing bank slides, loss of vegetation on the bank of rivers, disturbs the aquatic life as well as affects agriculture. So there is an immediate need to control the sand quarrying and provide a sustainable replacement of river sand. Properties of aggregate affect the durability and performance of concrete, so fine aggregate is an essential component of concrete.

Manufactured sand in concrete not only acts as replacement for concrete but also leads to the development of eco-friendly construction as well as reduction in cost of construction. Manufactured Sand is nothing but artificial sand made from crushing of rock or granite for construction purposes in cement or concrete. M sand differs from natural river sand in its physical and mineralogical properties.

Advantages of Manufactured Sand

- M Sand has higher Fineness Modules Index compared to the natural river sand, which gives good workability for concrete.
- M sand is free from silt and clay particles which offer better abrasion resistance, higher unit weight and lower permeability.
- Less disruptive to the environment, as it reduces sand mining from river beds.
- Perfect grading and cubical shape of M Sand gives high strength and great durability to concrete.
- More cost-effective than river sand due to low transportation cost and consistency in availability.

Disadvantages of Manufactured Sand

- Due to its smooth and angular textures, leads to more water and cement requirement to achieve the expected workability, thereby increase in overall costs.
- If the M Sand contains a large number of micro fine particles, it

can affect the strength and workability of concrete.

Geosynthetics

Geosynthetics are a type of plastic polymer products used in geotechnical engineering that improves engineering performance while reducing costs.

Figure 8.4 Different types of Geosynthetics.

This includes geotextiles, geogrids, geomembranes, geocells, and geocomposites. The synthetic nature of the products makes them suitable for use in the ground where high levels of durability are required; their main functions include drainage, filtration, reinforcement, separation, and containment. Geosynthetics are available in a wide range of forms and materials, each to suit a slightly different end-use, although they are frequently used together. These products have a wide range of applications and are currently used in many civil and geotechnical engineering applications including roads, airfields, railroads, embankments, piled embankments, retaining structures, reservoirs, canals, dams, landfills, bank protection and coastal engineering.

Chapter 9

Equilibrium of force systems & support reactions in beams

9.1 Types of Force Systems

There are various types of force systems that can exist depending on the arrangement and characteristics of the forces acting on an object. Some of the commonly encountered types of force systems include:

Concurrent Force System

In a concurrent force system, all the forces have their lines of action intersecting at a single point. The resultant force of a concurrent force system can be determined by finding the vector sum of all the individual forces. Examples of concurrent force systems include the forces acting on a pin-jointed structure or the forces exerted by a group of ropes attached to a common point.

Non-Concurrent Force System

In a non-concurrent force system, the lines of action of the forces do not intersect at a single point. This type of force system requires additional considerations such as moments and couples to determine the overall effect on the object. An example of a non-concurrent force system is the forces acting on a beam or a truss structure.

Coplanar Force System

A coplanar force system consists of forces that lie in the same plane. These forces can be concurrent or non-concurrent. Coplanar force systems are commonly encountered in two-dimensional problems and are often analyzed using methods such as vector addition or graphical techniques like the method of polygons.

Collinear Force System

In a collinear force system, all the forces have the same line of action or are parallel to each other. The resultant force in a collinear force system is obtained by algebraically summing the forces along the line of action. Examples include forces acting along a straight rod or a cable subjected to tension or compression.

Parallel Force System

A parallel force system consists of forces that have parallel lines of action. While the forces in a parallel force system do not necessarily have to be collinear, they share the same direction or are parallel to each other. Parallel force systems commonly arise in situations involving distributed loads, such as the weight distribution on a beam or the hydrostatic pressure distribution on a submerged surface.

9.2 Applications of System of Forces

System of forces find numerous applications across various fields of engineering and physics. Here are some notable applications:

Structural Engineering: In structural engineering, the analysis of system of forces is fundamental to designing safe and stable structures.

Mechanical Engineering: System of forces analysis plays a crucial role in mechanical engineering. It helps in the design of machines and mechanisms by considering the forces acting on various components.

Statics and Dynamics: The study of system of forces forms the basis of statics and dynamics, which are branches of mechanics. Statics deals with the equilibrium of objects under the influence of forces, while dynamics involves the motion and acceleration of objects.

Civil Engineering: In civil engineering, system of forces analysis is applied to various structural systems, such as bridges, dams, and retaining walls.

Aerospace Engineering: System of forces analysis is vital in aerospace engineering for designing and analyzing aircraft and spacecraft. The forces acting on the vehicles, such as lift, drag, thrust, and weight, are considered to ensure proper flight characteristics, stability, and control.

State of Equilibrium

Equilibrium means "no acceleration." Because a force is a "push" or "pull" applied to a body, Equilibrium requires that the sum of all forces acting on it be zero.

F = m * a, according to Newton's second law.

If a = 0, F must also be zero.

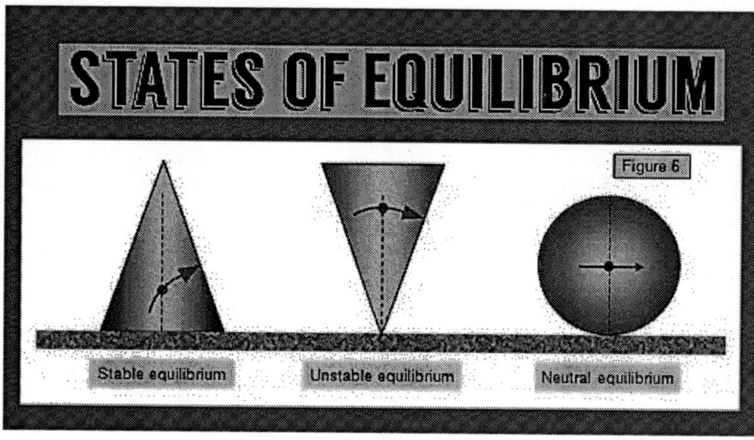

Figure 9.1 Types of Equilibrium

9.3 Concurrent Forces Equilibrium

The first step in comprehending the Equilibrium of a concurrent force system is understanding what Equilibrium is. Equilibrium is the state of the body in which all of the forces operating on it are in balance. This cancels out, resulting in a net impact of zero.

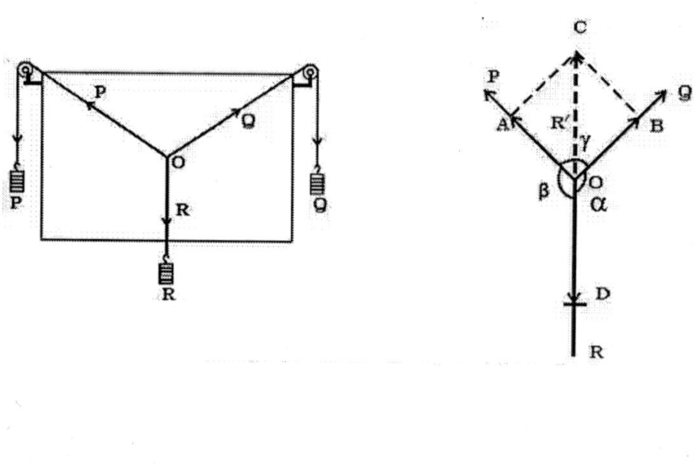

Figure 9.2 Concurrent Forces

Forces that cross at a common location are referred to as concurrent forces. We can combine them as vectors when dealing with concurrent forces to yield the net resultant. It's important to note that its acceleration is zero when a body is at Equilibrium. This is one of the equilibrium conditions for concurrent forces.

Various Kinds of Concurrent Forces Equilibrium

There are two forms of Equilibrium: Static and Dynamic.

Static Equilibrium

This is the form of Equilibrium in which the sum of all forces acting on the body is zero, i.e., the body's net acceleration is zero, and its velocity is likewise zero. It denotes that the body is in a state of rest. When a body is at rest and its net acceleration is zero, it is said to be in static Equilibrium.

Example - Assume a slab is laying on the ground and two 5 N forces are operating from opposite directions. The forces would balance one another out, leading to zero net force on the slab. The slab will be in static Equilibrium since it is at rest.

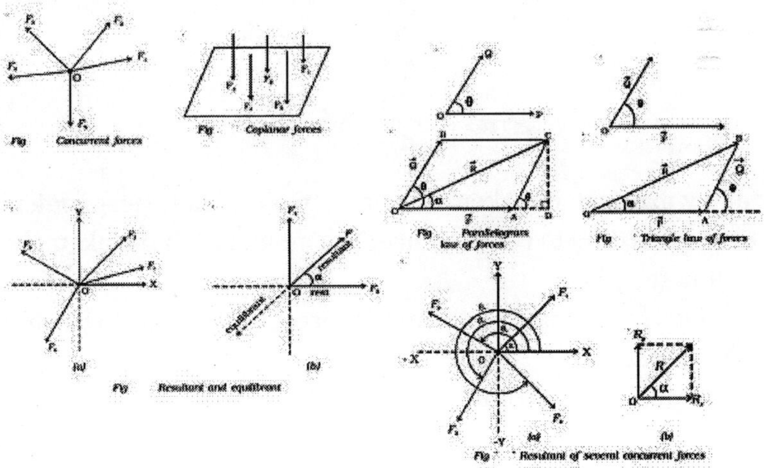

Figure 9.3 Types of forces and resolution of forces

Dynamic Equilibrium

This kind of Equilibrium is when the sum of all forces acting on the body is zero, i.e., the body's net acceleration is zero, but its velocity is not zero. It indicates that the object is traveling at a steady speed. The body is considered in dynamic Equilibrium if the net force exerted on it is zero and still traveling at a constant velocity.

Example - A popular illustration of dynamic Equilibrium is a block tied to a spring under the effect of Simple Harmonic Motion (S.H.M.). The net force exerted on the block is zero at the mean position, but the block's velocity is greatest, indicating that the block is in dynamic Equilibrium at that point.

Equilibrium of Coplanar Non-Concurrent Force System

- As discussed in Chapter 3, the coplanar non-concurrent force systems can be reduced to a resultant force and a couple moment. Hence the conditions for equilibrium are

$$\sum F_x = 0$$

$$\sum F_y = 0$$

$$\text{and} \sum M = 0$$

- x and y axes can be taken along any two mutually perpendicular directions according to convenience. The moment can be taken about any point in the plane.
- Problems involving equilibrium of three forces can also be solved using this method.
- For beams, the distributed loads have to be converted to point load

Special Contact Points

We must include any forces that cross the body's border (outline) while designing the Free Body Diagram. In addition, some touch places provide unique information about the force's direction and location.

Cable Joint: The force must pass through the cable's centreline.

The force must be perpendicular to the surface in a frictionless joint.

Wheels and rollers: The force must be transmitted through the axle's center. Because free-running wheels have no **friction**, force is applied perpendicular to the surface, and all forces flow through the axle's center.

Pulleys: The cable tension is the same on both sides of the pulley, and all forces may be directed via the axle's center.

Friction is a force that may be applied in any direction.

The force can be applied in either direction at a pin joint.

Support Reactions are another name for contact points.

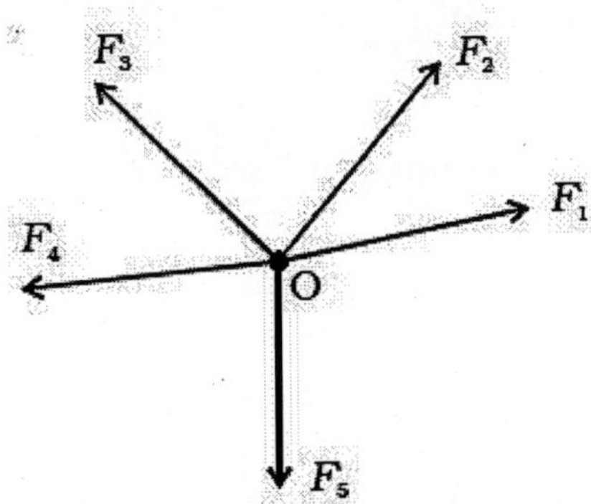

Figure 9.4 Concurrent Forces

Coplanar Forces in Equilibrium

Forces that act on the same plane are known as coplanar forces. For Coplanar Forces to be in balance, several requirements must be met. These are some of them:

- The total force must equal zero.
- The total of the moments of forces around a point in the clockwise direction equals the sum of the moments of forces around the same point in the anticlockwise direction.

- Coplanar forces come in a variety of shapes and sizes. The forces in one plane must cross in concurrent coplanar forces of Equilibrium at one point. Graphical and algebraic approaches can be used to compute it.
- Non-concurrent and parallel coplanar forces are the other forms of coplanar forces. The vectors of non-concurrent coplanar forces do not meet at a single location and are not parallel. The forces in parallel coplanar forces are all parallel to one another.

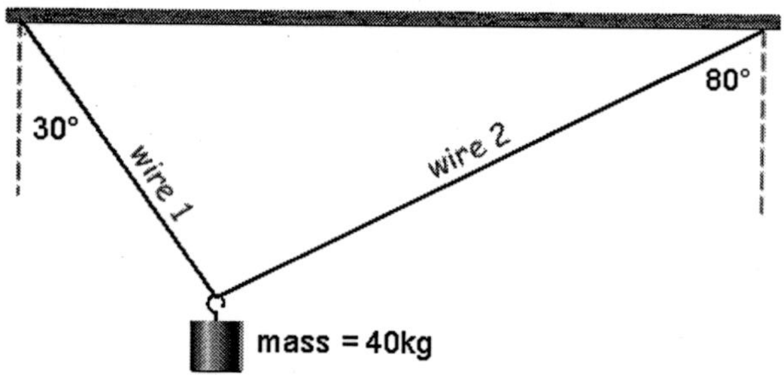

Figure 9.5 Coplanar Forces in Equilibrium

9.4 The Elements of Coplanar Force Resolution

There are many ways in which forces can be manipulated. It is often easier to work with a large, complicated system of forces by reducing it to an ever decreasing number of smaller problems. This is called the "resolution" of forces or force systems. This is one way to simplify what may otherwise seem to be an impossible system of forces acting on a body. Certain systems of forces are easier to resolve than others. **Coplanar** force systems have all the forces acting in in one plane. They may be concurrent, parallel, non-concurrent or non-parallel. All of these systems can be resolved by using graphic statics or algebra.

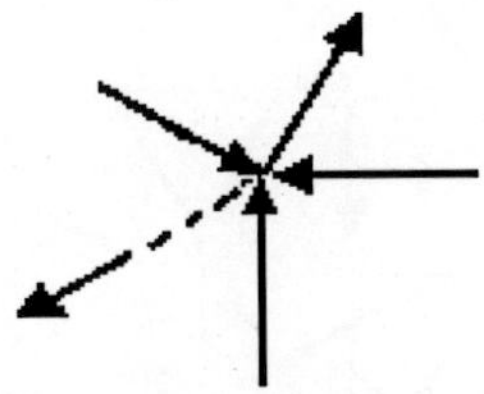

Concurrent

A **concurrent** coplanar force system is a system of two or more forces whose lines of action ALL intersect at a common point. However, all of the individual vectors might not acutally be in contact with the common point. These are the most simple force systems to resolve with any one of many graphical or algebraic options.

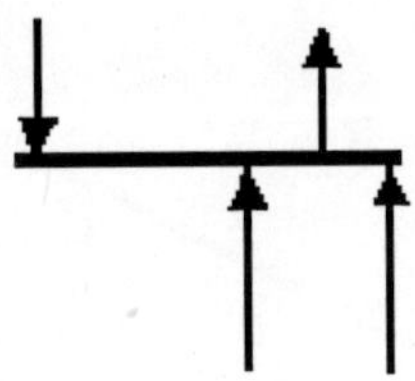

Parallel

A **parallel** coplanar force system consists of two or more forces whose lines of action are ALL parallel. This is commonly the situation when simple beams are analyzed under gravity loads. These can be solved graphically, but are combined most easily using algebraic methods.

The last illustration is of a **non-concurrent and non-parallel** system. This consists of a number of vectors that do not meet at a single point and none of them are parallel. These systems are essentially a jumble of forces and take considerable care to resolve.

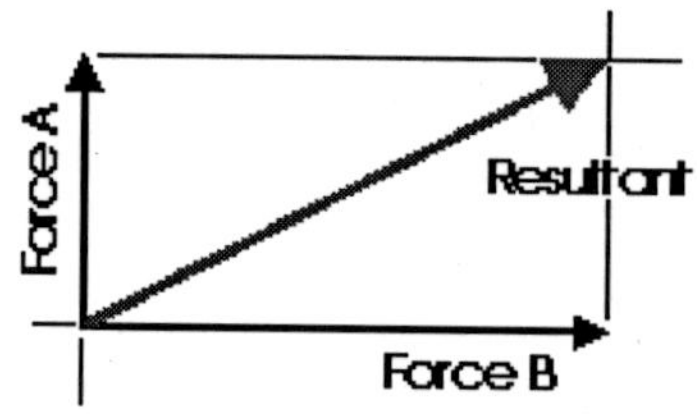

Almost any system of known forces can be resolved into a single force called a resultant force or simply a Resultant. The resultant is a representative force which has the same effect on the body as the group of forces it replaces. (A couple is an exception to this) It, as one single force, can represent any number of forces and is very useful when resolving multiple groups of forces. One can progressively resolve pairs or small groups of forces into resultants. Then another resultant of the resultants can be found and so on until all of the forces have been combined into one force. This is one way to save time with the tedious "bookkeeping" involved with a large number of individual forces. Resultants can be determined both graphically and algebraically. The graphical methods that will be discussed in Lecture 8 are the

Parallelogram Method and the Triangle Method. It is important to note that for any given system of forces, there is only one resultant.

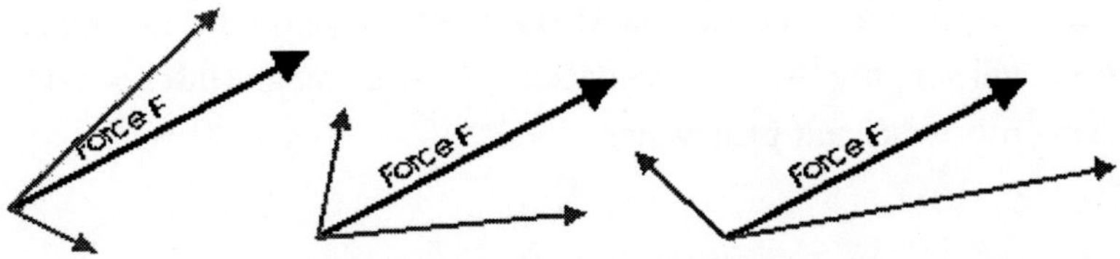

It is often convenient to decompose a single force into two distinct forces. These forces, when acting together, have the same external effect on a body as the original force. They are known as **components**. Finding the components of a force can be viewed as the converse of finding a resultant. There are an infinate number of components to any single force. And, the correct choice of the pair to represent a force depends upon the most convenient geometry. For simplicity, the most convenient is often the coordinate axis of a structure.

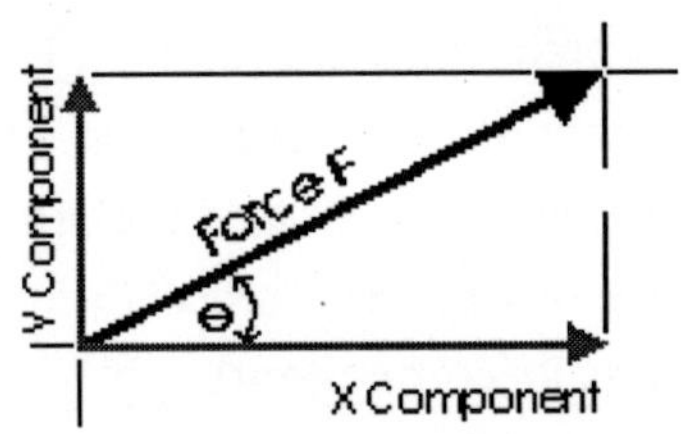

This diagram illustrates a pair of components that correspond with the X and Y axis. These are known as the **rectangular components** of a force. Rectangular components can be thought of as the two sides of a right angle which are at ninety degrees to each other. The resultant of these components is the hypotenuse of the triangle. The rectangular components for any force can be found with trigonometrical relationships: F_x = F cos theta, F_y = F sin theta (where theta is the angle). There are a few geometric relationships that seem common in general building practice in North America. These relationships relate to roof pitches, stair pitches, and common slopes or relationships between truss members. Some of these are triangles with sides of ratios of 3-4-5, 1-2-sqrt3, 1-1-sqrt2, 5-12-13 or 8-15-17. Commiting the first three to memory will simplify the determination of vector magnitudes when resolving more difficult problems.

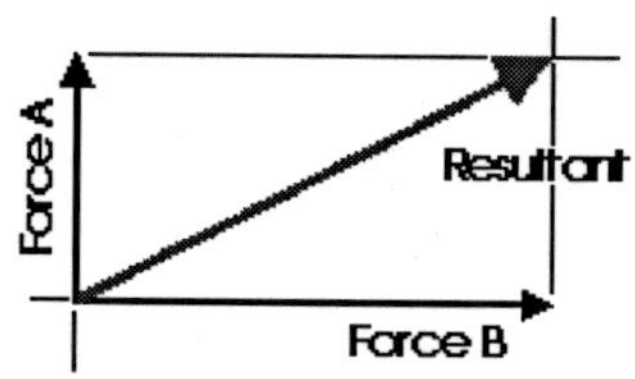

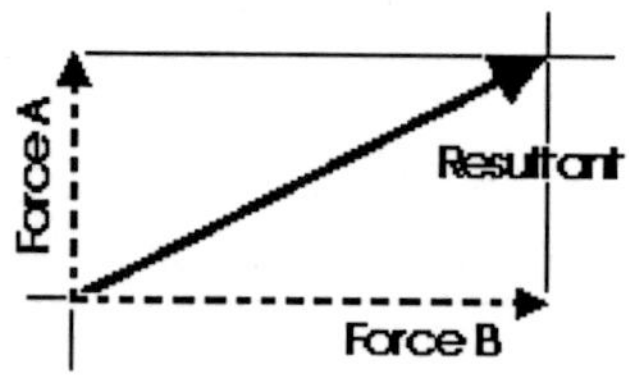

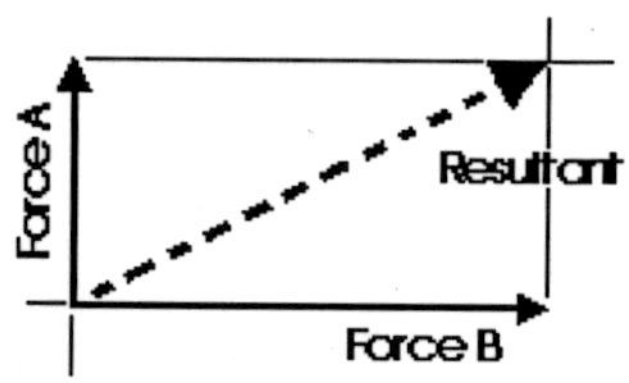

When forces are being represented as vectors, it is important to should show a clear distinction between a resultant and its components. The resultant could be shown with color or as a dashed line and the components as solid lines, or vice versa. NEVER represent the resultant in the same graphic way as its components.

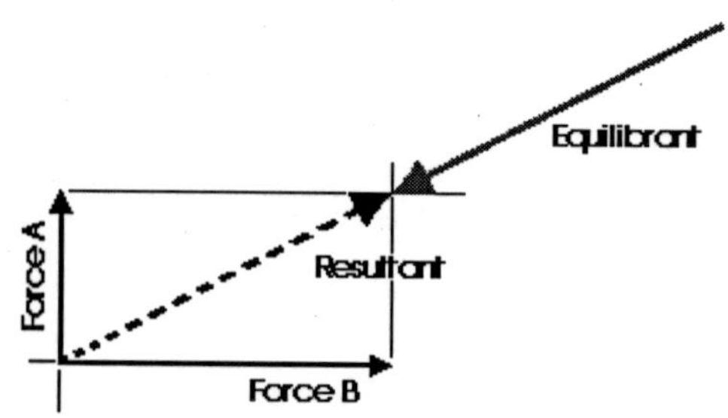

Any concurrent set of forces, not in equilibrium, can be put into a state of equilibrium by a single force. This force is called the **Equilibrant.** It is equal in magnitude, opposite in sense and co-linear with the resultant. When this force is added to the force system, the sum of all of the forces is equal to zero. A non-concurrent or a parallel force system can actually be in equilibrium with respect to all of the forces, but not be in equilibrium with respect to moments.

Graphic Statics and graphical methods of force resolution were developed before the turn of the century by Karl Culmann. They were the only methods of structural analysis for many years. These methods can help to develop an intuitive understanding of the action of the forces. Today, the Algebraic Method is considered to be more applicable to structural design. Despite this, graphical methods are a very easy way to get a quick answer for a structural design problem and can aid in the determination of structural form.

The following is an example of how an understanding of structural action and force resolution can aid in the interpretation of structural form. The illustration on the left is the Saltash Bridge, design and built by the Engineer Brunel in 1859. Notice the form of the structural elements. The illustration on the right is a retouched imaged to bring the bridge to a state which is akin to a pure suspension bridge. However, there are some issues with this design that need to be clearly

Conditions of Static Equilibrium of Concurrent Forces

In the y-direction or vertical, the total of all forces is zero. $\Sigma Fy=0$ or $\Sigma FV=0$

- In the x-direction or horizontal, the total of all forces is zero. $\Sigma Fx=0$ or $\Sigma FH=0$

Things to Remember

When two forces are equal and oppositely directed, they are in Equilibrium.

- The vectors are not parallel to each other and do not meet at a location for non-concurrent coplanar forces.
- Three parallel coplanar forces are in balance.

- It is impossible to achieve Equilibrium when only a single force is applied to the body.
- When three or more equilibrium forces are coupled head-to-tail, they create a close polygon.

Types of Supports and Their Characteristics

The type of support provided for a structure is important in ensuring its stability. Supports connect the member to the ground or to some other parts of the structure. It is assumed that the student is already familiar with several types of supports for rigid bodies, as this was introduced in the statics course.

Pin or Hinge Support

A pin support allows rotation about any axis but prevents movement in the horizontal and vertical directions

3.2.2 Roller Support

A roller support allows rotation about any axis and translation (horizontal movement) in any direction parallel to the surface on which it rests. It restrains the structure from movement in a vertical direction.

3.2.3 Rocker Support

The characteristics of a rocker support are like those of the roller support.

3.2.4 Link

A link has two hinges, one at each end. It permits movement in all direction, except in a direction parallel to its longitudinal axis, which passes through the two hinges. In other words, the reaction force of a link is in the direction of the link, along its longitudinal axis.

Fixed Support

A fixed support offers a constraint against rotation in any direction, and it prevents movement in both horizontal and vertical directions.

Determinacy and Stability of Beams and Frames

Prior to the choice of an analytical method, it is important to establish the determinacy and stability of a structure. A determinate structure is one whose unknown external reaction or internal members can be determined using only the conditions of equilibrium.

An indeterminate structure is one whose unknown forces cannot be determined by the conditions of static equilibrium alone and will require, in addition, a consideration of the compatibility conditions of different parts of the structure for its complete analysis. Furthermore, structures must be stable to be able to serve their desirable functions. A structure is considered stable if it maintains its geometrical shape when subjected to external forces.

3.3.1 Formulations for Stability and Determinacy of Beams and Frames

The conditions of determinacy, indeterminacy, and instability of beams and frames can be stated as follows:

$$\begin{aligned} 3m + r &< 3j + C \quad \text{Structure is statically unstable} \\ 3m + r &= 3j + C \quad \text{Structure is statically determinate} \\ 3m + r &> 3j + C \quad \text{Structure is statically indeterminate} \end{aligned} \tag{3.3}$$

where

r = number of support reactions.

C = equations of condition (two equations for one internal roller and one equation for each internal pin).

m = number of members.

j = number of joints.

9.5 LAMI'S THEOREM

Statement:

"If three force acting on a particle are in equilibrium, then each force is directly

proportional to the sine of the angle between the other two forces".

Explanation:

Let P, Q and R three forces acting at a point keeping it in equilibrium, If α, β and γ are the

angles opposite to each of them respectively.

P/sinα = Q/sinβ = R/sinγ

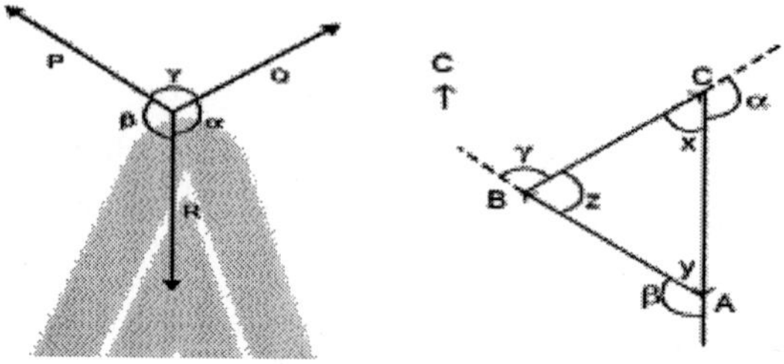

Figure 9.5 Lames Theorem

This law is a direct consequence of the triangle law. Since the forces in equilibrium, they

can be represented by the sides of the triangle ABC taken in order. A general property of

any triangle is that each side is proportional to the sine of the angle opposite to it. Thus in

the triangle ABC drawn with sides parallel to the forces P, Q and R.

AB/sinx = BC/siny =AC/sinz

Here x, y and z are the angle if the triangle ABC. But by the triangle law of forces, the

sides of the triangle are proportional to the respective force. From the above figure

$\sin x = \sin (180-\alpha) = \sin \alpha$

$\sin y = \sin (180-\beta) = \sin \beta$

$\sin z = \sin (180- \gamma) = \sin \gamma$

Hence,

$P/\sin\alpha = Q/\sin\beta = R/\sin\gamma$ = constant

Thus, if three forces acting on a particle are in equilibrium, each force is proportional to the sine of the angle between the other two forces.

Significance of Lamis theorem:

 It is used to find out the unknown forces where three concurrent and coplanar force

acting on a body and the body should be in the state of equilibrium.

Types of supports

There are 3 types of supports

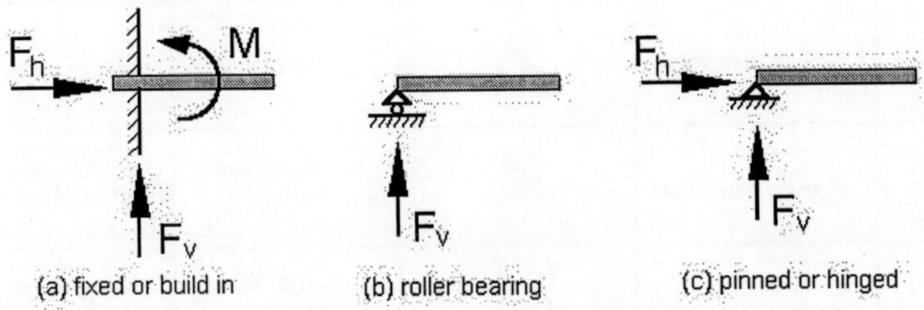

Figure 9.6 Types of forces

1. Roller support:

If a beam is supported on a roller support then it can't in vertical direction but it can move in horizontal direction and can have rotational motion about that support.

Hence total support reactions are 1 i.e in vertical direction.

2. Hinge support / Pinned support

Similarly hinge support restricts vertical and horizontal motion but beam is free to perform rotational motion.

Hence it has 2 support reactions i.e horizontal and vertical

3. Fixed / Build in support:

Here the support restricts all kinds of motion. The can't perform horizontal, vertical and rotational motion.

Hence it has 3 support reactions i.e horizontal, vertical and rotational reactions.

S.no	Types of Support	Representation by	Reaction Force	Resisting Load
1.	**Roller Support**		Vertical	Vertical loads
2.	**Pinned Support**		Horizontal and vertical	Vertical and horizontal loads
3.	**Fixed Support**		Horizontal, vertical and moments	All types of loads Horizontal, vertical and Moments
4.	**Simple Support**	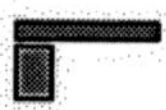	Vertical	Vertical loads

9.6 Types of Beams

A beam is a structural member designed to support various loading applied at points along the member.

The forces / loads applied to a beam are counteracted by the supports which are called support reactions

Based on support conditions the beams are called in the following types.

1.Simply supported Beam :

a beam supported on the ends which are free to rotate and have no moment resistance.

2.Cantilever Beam :

a projecting beam fixed only at one end

3.Over hanging Beam :

a simple beam extending beyond its support on one end.

4.Double overhanging Beam :

a simple beam with both ends extending beyond its supports on both ends.

5.Continuous Beam :

a beam extending over more than two supports.

6.Fixed Beam :

a beam supported on both ends fixed and restrained from rotation.

7.Propped Cantilever Beam :

a beam supported on fixed end and other end is supported by a simple support / prop.

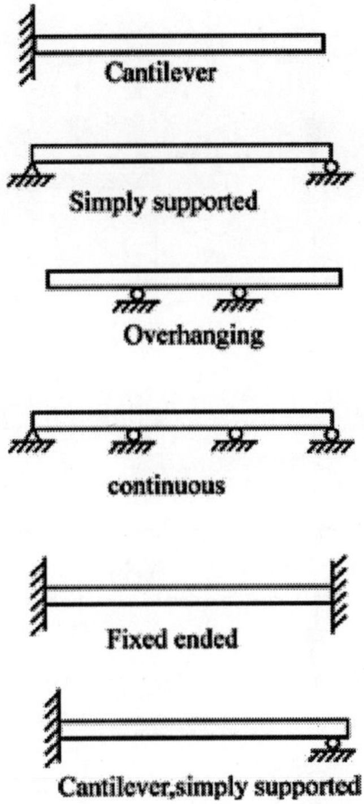

Figure 9.7 Types of Beams

9.7 Types of Loads

The following are the different types of loads applied to beams.

1. Concentrated or Point Load: Load is assumed to act at a point.

2.Uniformly Distributed Load (U.D.L.): Load spread along the length of the Beam uniformly.

3.Uniformly Varying Load: Load spread along the length of the Beam, Rate of varying loading point to point.

4.Applied couple or pure moment: a **couple** refers to two parallel forces that are equal in magnitude, opposite in direction and its resultant is in the form of a pure moment acting at the connected point.

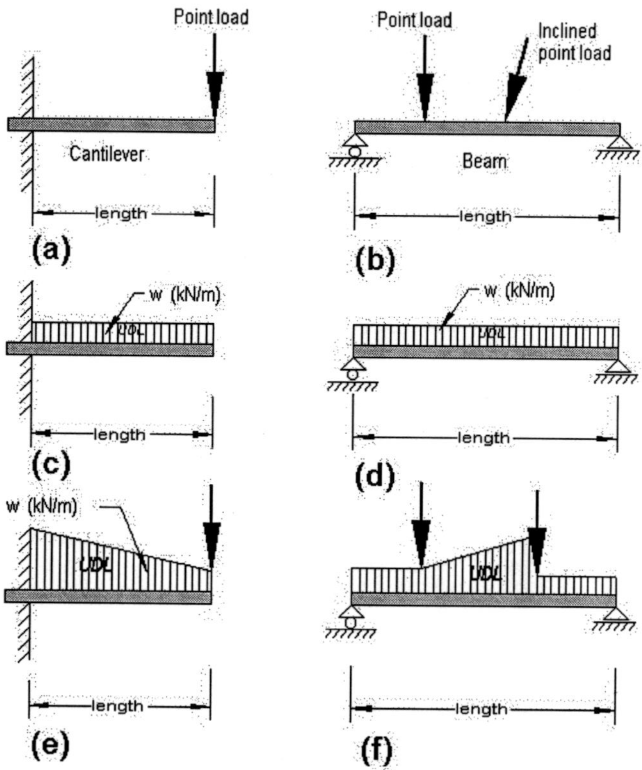

Figure 9.8 Types of Loads on Beams

Support Reactions in Beams

The Support Reactions in Beams will find out by using the three static equilibrium equations.

Equations of Static Equilibrium

$\Sigma F_x = 0 \quad \Sigma F_y = 0 \qquad \Sigma M_P = 0$

F_x = force in the x-direction
F_y = force in the y-direction
M_P = moment about point P

The support reactions are less than three we can determine the values by using the three static equilibrium equations. these beams are called **Statically determinate beams.**

If the unknown reactions are more than three the beam is considered as **Statically indeterminate beam.**

Note :

In the simplification of the distributed loading system, the total load is converted to an equivalent point load. The net equivalent point load acts at the centroid of the loading body.

a.If the loading system is in the form of the triangle (distributed uniformly varying load) then the net force of the loading will act at centroid of triangle. The net point load will be the area of the triangular loading.

b. If the loading system is in the form of the rectangular (uniformly distributed load) then the net force of the loading will act at centroid of rectangular. The net point load will be the area of the rectangular loading.

c. If the loading system is in the form of parabola, the net force will act at the centroid of the parabola. Whether it be a parabola or the cubic curve the centroid is the only point at which the net force act.

When solving for reactions, the following steps are recommended:

1. Draw the beam free body diagram
2. Replace the uniform distributed load (if any) with the equivalent point load
3. Solve $\Sigma M_A = 0$ (sum of moments about support A). This will give you R_B (reaction at support B).

4. Solve $\Sigma M_B = 0$. This will give you R_A.
5. Using R_A and R_B found at steps 3 and 4 check if $\Sigma V = 0$ (sum of all vertical forces) is satisfied.
1. Note that steps 4 and 5 can be reversed.
2. For a cantilever beam use $\Sigma V = 0$ to find the vertical reaction at the wall and $\Sigma M_{wall} = 0$ to find the moment reaction at the wall. There is no other equation to validate your results.

Referencess

1. Mechanical Engineering Science -J. L. Gwyther, B.Sc.(Eng.), B.Sc.(Econ.), C.Eng., F.I.Mech.E. Vice-Principal, College of Technology, Letch worth W. D. Brown, M.Sc, C.Eng., M.I.Mech.E., M.T.I. Sometime Principal Scientific Officer, Ministry of Aircraft Production G. Williams, B.Sc.(Eng.), C.Eng., M.I.C.E., M.I.Struct.E. Deputy Engineer, Severn River Authority. Heinemann Educational Books Ltd London.

2. Mechanical Engineering Science Paperback – 1 March 1999, by J. Hannah (Author), M. J. Hillier (Author). Third Edition, Prentice Hall Publisher.

3. Applied Mechanics Paperback – 19 June 1995,by J. Hannah (Author), M. J. Hillier (Author), M.J. Hillier (Author), Third Edition , Prentice Hall Publisher.

4. https://make.3dexperience.3ds.com/processes/additive

5. LECTURE NOTE On CIVIL ENGINEERING MATERIALS and CONSTRUCTION COURSE CODE: BCE03002: 3.0.0 (CR 03) Third Semester, B Tech, Civil Engineering Dr. S K Panigrahi Associate Professor Dept. of Civil Engg , VSSUT BURLA

6. https://satpalda.com/gis-in-civil-engineering- applications

7. Elements of Mechanical Engineering, Dr. K R Gopala Krishana, Subash Publications.